UNDERSTANDING
QUANTUM PHYSICS

Umberto Piacquadio

Not only is the Universe stranger than we think, it is stranger than we can think

(Werner Heisenberg)

INDEX

Preface

After the formulations of Newton's and Maxwell's theories, it seemed that nothing else could scratch the path taken in the Physical description of natural phenomena.

The principle of causality well described electromagnetic phenomena and the motion of celestial bodies in the sign of determinism.

With the formulation of the Theory of Relativity Einstein continues a path of describing phenomena in motion even at speeds comparable to those of light.

The change began with the deepening and studies at the atomic level where, Planck, Bohr, Einstein, abandoned the concept of continuity in favor of introducing the quantization of matter, while continuing, however, to interpret phenomena always with a classical foundation.

It will take the input of the young Heinsenberg and Schoidenger to finally abandon the classical interpretation in the face of the mind-blowing new quantum interpretation.

The subsequent formalization of quantum mechanics based on noncommutative algebra introduced by the young Dirac continues the path toward the ultimate use of quantum theory in the microscopic world.

With quantum physics, a whole new way of behavior of matter and light is discovered in the realm of the microcosm.

An atom no longer consists of electrons orbiting like planets, no longer possessing a definite trajectory with certain values of velocity and position.

Quantum physics describes the microcosm by preferring an evolution of nature toward disorder and uncertainty, rather than the determinism established by cause-and-effect, dominant in classical physics.

Particles can transmit instantaneous information to each other, beyond the limit of the speed of light imposed by Einstein with the formulation of the Theory of Special Relativity.

Understanding the behavior of matter at the atomic level will lead you to important reflections, and to think that nothing around us can be interpreted with deterministic rationality alone anymore.

Hundreds of books can be found to understand the fabulous world of quantum physics, but most of the time they are either too popular in nature or are treated at the university level.

With the present exposition, however, I have tried to cover the topics so that the concepts are mainly understood, but without neglecting the rigorous mathematical formulas and demonstrations, with sufficient language to be interpreted with mathematical and physical skills learned in a high school.

The present text does not claim the prerogative of being exhaustive in the interpretation of quantum theory, but it is certainly useful for acquiring notions in order to be able to understand in a scientific view the texts on the market of a popular nature on the subject and in any case create excellent prerequisites for future in-depth studies of a university nature.

I would like to thank all those who have been close to me during the writing of this treatise, and with the hope that I have set up the work in such a way that it will be useful to all those who approach the study of the fascinating world of Quantum Physics, I am grateful in advance to those who would like to propose improvements or any suggestions.

1. ATOMIC HISTORY.

The history of the constitution of matter, from an atomic point of view, finds its origins as early as 1500 to 500 B.C. in India, where philosophical schools, in order to

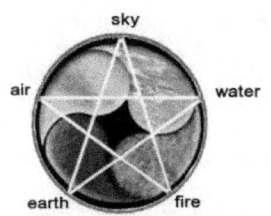

justify the existence of matter, identified five fundamental essences as essential components: Fire-Earth-Air-Water-Sky.

In addition to the five substances, they completed the constitution of the known, additional four external senses: Space, Time, Mind and finally Ego.

Matter thus composed was always divisible into a finite number of particles, according to the fundamental essences and external senses.

Next, in ancient Greece, Leucippus of Miletus first used the term Atom (ἄτομος), meaning indivisible, and his student Democritus

of Abdera, in 460-360 B.C., made the term famous. In the latter interpretation, the atom was considered an indivisible particle that formed all known matter.

Everything else was filled by emptiness, like the same place where atoms come together in a concept of eternity, through their birth, death and rebirth.

Divisibility to infinity, remained valid only in the logical-mathematical field, but not for matter, which in a process of infinite divisibility would dissolve into nothingness, until an unworkable concept of non-matter was reached.

Later, in 384-322 B.C. he advances Aristotle's theory, where matter is understood as infinitely divisible, thus abandoning the concept of the existence of the indivisible atom.

For Aristotle, the magnitude of a body's motion certainly depended on its weight, but mainly on the medium it passed through, the composition of which quenched its velocity.

There could be no vacuum, which represented nothingness, as it would cause a body to reach infinite speed, against common sense.

Contrary to Democritus' consideration that there must be a vacuum among the indivisible atoms, Aristotle maintained that matter must be continuous and infinitely divisible throughout its formation of earthly existence, under the direction and control of a divine creator entity.

In the course of its continuous division, matter would change properties and substances would be transformed into others in order to encompass every possible entity of the known.

The Middle Ages (A.D. 400-1400), on the other hand, represented the darkest period of atomistic philosophy because of the proliferation of esoteric practices, which were closer to magic than to science. It was during this period that alchemy gained prominence, remembered for its characteristic of wanting to transmute vile metals into gold.

This conception is fortunately abandoned with the onset of the "Renaissance" and continues into the "Baroque" era as a period of scientific revolution, where a clear separation between science and religion begins to take shape.

The latter era marks the beginning of the experimental method in the sciences by the great father of modern science, Italian and physicist, astronomer, philosopher, mathematician and academic, Galileo Galilei.

For the first scientific concept of the atom, it would have to wait until the year 1807 for the theories of the English chemist, physicist, meteorologist and teacher, John Dalton, who was also famous for giving

rise to the famous term "color blindness," from which he was afflicted.

Dalton, through the decomposition of water into its components Oxygen and Hydrogen, analyzing their proportions, understood that in a chemical reaction the atoms remain unchanged in number and mass, respecting the principle of conservation of mass and combining in whole number proportions.

In the case of water, two hydrogen (H) atoms combined with one oxygen (O) atom, which we now write with the more famous chemical formula H_2O.

In about 1823, Lorenzo Romano Amedeo Carlo Avogadro, an Italian chemist and physicist, introduced the concept of a molecule as being composed of atoms, and derived experimentally that equal volumes of gas, even different ones, at the same temperature and pressure contain the same number of molecules and thus atoms.

In that hypothesis, by weighing volumes of different gases, the proportions between the different atomic masses could be derived.

Further research by Coulomb-Faraday- Maxwell between 1780-1830, on the mathematical theories of electricity and magnetism (Charles Augustin De Coulomb), on the correlation of mass and quantity of electric charge (Michael Faraday), and on the unification of electromagnetic theory (James Clerk Maxwell) prepared the field for the advent of atomic physics.

In 1869, D.I. Mendeleev, a Russian chemist, devised an order in chemistry by arranging the chemical elements in order of increasing atomic weight and in periods, forming a periodic table, the forerunner of today's periodic table.

Mendeleev's work is exceptional in that in those days the composition of the atom was not yet known, let alone the existence of electrons, yet he still managed to group the elements by same chemical properties.

In Mendeleev's periodic table, the elements were arranged in rows and columns, in order of atomic mass, in an appropriate arrangement by rows and columns when the characteristics of the elements began to repeat.

To said table he made some changes and unknowingly performed an order by atomic number (number of protons).

ОПЫТЪ СИСТЕМЫ ЭЛЕМЕНТОВЪ.

ОСНОВАННОЙ НА ИХЪ АТОМНОМЪ ВѢСѢ И ХИМИЧЕСКОМЪ СХОДСТВѢ.

```
                        Ti = 50   Zr = 90    ? = 180.
                        V = 51    Nb = 94    Ta = 182.
                        Cr = 52   Mo = 96    W = 186.
                        Mn = 55   Rh = 104,4 Pt = 197,4.
                        Fe = 56   Rn = 104,4 Ir = 198.
                        Ni = Co = 59  Pl = 106,6 O = 199.
H = 1                   Cu = 63,4 Ag = 108   Hg = 200.
           Be = 9,4 Mg = 24  Zn = 65,2 Cd = 112
           B = 11   Al = 27,4 ? = 68   Ur = 116   Au = 197?
           C = 12   Si = 28   ? = 70   Sn = 118
           N = 14   P = 31    As = 75  Sb = 122   Bi = 210?
           O = 16   S = 32    Se = 79,4 Te = 128?
           F = 19   Cl = 35,6 Br = 80   I = 127
Li = 7 Na = 23      K = 39    Rb = 85,4 Cs = 133  Tl = 204.
                    Ca = 40   Sr = 87,6 Ba = 137  Pb = 207.
                    ? = 45    Ce = 92
                    ?Er = 56  La = 94
                    ?Yt = 60  Di = 95
                    ?In = 75,6 Th = 118?
```

Д. Менделѣевъ

As knowledge of the atom progressed and new elements were discovered, the periodic table was appropriately updated.

The current configuration of the modern periodic table consists of a diagram in which certain chemical and physical characteristics of the chemical elements are shown, and the elements are ordered on the basis of their atomic number Z (number of protons = number of electrons) from left to right and from top to bottom (like the order of writing), appropriately grouped by similar characteristics.

Periodic Table of the Elements

Legend: metals, nonmetals, metalloids

Key:
Atomic Number — 6
Symbol — C
Name — Carbon
Average Atomic Mass — 12.011

1	2	3	4	5	6	7	8	9	10	11	12	13	14	15	16	17	18
1 H 1.008																	2 He 4.003
3 Li 6.94	4 Be 9.012											5 B 10.81	6 C 12.011	7 N 14.007	8 O 15.999	9 F 18.998	10 Ne 20.180
11 Na 22.990	12 Mg 24.305											13 Al 26.982	14 Si 28.085	15 P 30.974	16 S 32.06	17 Cl 35.45	18 Ar 39.948
19 K 39.098	20 Ca 40.078	21 Sc 44.956	22 Ti 47.867	23 V 50.942	24 Cr 51.996	25 Mn 54.938	26 Fe 55.845	27 Co 58.933	28 Ni 58.693	29 Cu 63.546	30 Zn 65.38	31 Ga 69.723	32 Ge 72.630	33 As 74.922	34 Se 78.971	35 Br 79.904	36 Kr 83.798
37 Rb 85.468	38 Sr 87.62	39 Y 88.906	40 Zr 91.224	41 Nb 92.906	42 Mo 95.95	43 Tc [97]	44 Ru 101.07	45 Rh 102.906	46 Pd 106.42	47 Ag 107.868	48 Cd 112.414	49 In 114.818	50 Sn 118.71	51 Sb 121.760	52 Te 127.60	53 I 126.904	54 Xe 131.293
55 Cs 132.905	56 Ba 137.327	71 Lu 174.967	72 Hf 178.49	73 Ta 180.948	74 W 183.84	75 Re 186.207	76 Os 190.23	77 Ir 192.217	78 Pt 195.084	79 Au 196.967	80 Hg 200.592	81 Tl 204.38	82 Pb 207.2	83 Bi 208.980	84 Po [209]	85 At [210]	86 Rn [222]
87 Fr [223]	88 Ra [226]	103 Lr [262]	104 Rf [267]	105 Db [270]	106 Sg [269]	107 Bh [270]	108 Hs [270]	109 Mt [278]	110 Ds [281]	111 Rg [281]	112 Cn [285]	113 Nh [286]	114 Fl [289]	115 Mc [289]	116 Lv [293]	117 Ts [293]	118 Og [294]

57 – 70 *
89 – 102 **

*Lanthanide series

57 La 138.905	58 Ce 140.116	59 Pr 140.908	60 Nd 144.242	61 Pm [145]	62 Sm 150.36	63 Eu 151.964	64 Gd 157.25	65 Tb 158.925	66 Dy 162.500	67 Ho 164.930	68 Er 167.259	69 Tm 168.934	70 Yb 173.045

**Actinide series

89 Ac [227]	90 Th 232.038	91 Pa 231.036	92 U 238.029	93 Np [237]	94 Pu [244]	95 Am [243]	96 Cm [247]	97 Bk [247]	98 Cf [251]	99 Es [252]	100 Fm [257]	101 Md [258]	102 No [259]

2. THE ATOMIC MODELS

2.1 THOMPSON AND RUTHERFORD

So far we have told of the results obtained by distinguished scientists, over the centuries, through elaborate theories formulated without knowledge of the mechanisms of the birth of electric charge. In particular, the correct composition of the atom and the existence of the electron were not yet known.

In 1897 British physicist Joseph John Thompson studied cathode rays in depth, thus leading to the discovery of the electron.

 Thompson was born in Cheetham Hill, Manchester on December 18, 1856, and at only 28 years of age was called to head one of the most famous research centers at the University of Cambridge, the Cavendish Laboratory, where among other things he became a lecturer endowed with extraordinary teaching skills. For several years he held the presidency of the Royal Society, that is, of England's highest academy. He died in Cambridge on August 30, 1940.

Cathode rays are luminescences that develop in a glass tube under vacuum or properly filled, as a result of an electrical source connecting to two plates: positive pole (anode) and negative pole (cathode).

The first cathode ray tube in history (Crookes tube) was made by William Crookes, in the early 1870s.

Applying an electric or magnetic field, he discovered that these were deflected, deducing that they could not be rays (electromagnetic waves), but rather negatively charged particles.

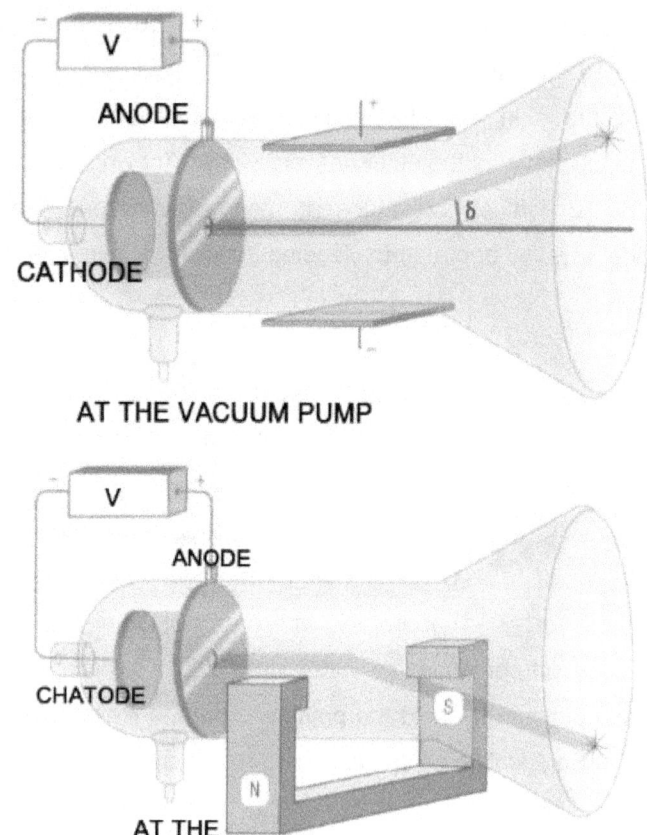

AT THE VACUUM PUMP

AT THE
VACUUM PUMP

Thompson, after discovering the consistency of the observed rays, through measuring the angle of deflection, was also able to derive the charge-to-mass ratio (q/m).

These small, thus identified, negatively charged particles came to be called "electrons."

The discovery of the electron constitutes the first real, experimentally supported discretization of matter.

In fact unlike Maxwell's earlier assumptions through the use of electric charge density to describe electrical phenomena, so that electric charge could take on any value of a continuous type, now electric charge could only take on multiple values of the elementary charge "*e.*"

Based on his experimental findings, the scientist, formulates the first modeling of the atom in history.

Thomson's atomic model is jokingly called the panettone atomic model. Just as the panettone has a distribution of raisins inside, so the atom should have had a uniform positive mass with the electrons distributed within it, all so that it would still have a neutral total charge, for such it had to be on the basis of experimental values.

 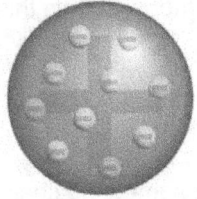

For that discovery in 1906 he received the Nobel Prize.

This model of the atom, however, is not long-lived.

It was New Zealand chemist and physicist Ernest Rutherford, who in 1909, at the same time as the discovery of "coulombic or Rutherford scattering," sanctioned the abandonment of the pancake atomic model for a new atomic model.

Ernest Rutherford, I Baron Rutherford of Nelson, was born in Spring Grove (now Brightwater), New Zealand, on August 30, 1871. He studied at Nelson College and Canterbury College, earning three degrees and two years of frontline research in electrical  technology. In 1895 he moved to England for postgraduate studies at the Cavendish Laboratory, University of Cambridge. During his investigation of radioactivity he coined the terms alpha and beta rays. In 1898 Rutherford was appointed to the chair of physics at McGill University, Canada, where he developed the work that won him the Nobel Prize in Chemistry in 1908. He had shown that radioactivity was the spontaneous disintegration of atoms. After noticing that in a sample of radioactive material he had a well-determined half-life, he devised a practical application of this phenomenon, using this constant rate of decay as a clock, to derive a determination of the actual age of the Earth, which turned out to be much older than most scientists of the time believed.

In 1907 he assumed the chair of physics at Victoria University of Manchester. There he discovered the existence of the atomic nucleus in atoms.   Later, while working with Niels Bohr, Rutherford made a proposal about the existence of neutral particles, the neutrons. In 1917 he returned to the Cavendish as Director. Under his direction, Nobel Prizes were awarded to James Chadwick for the discovery of the neutron, John Cockcroft and Ernest Walton for splitting the atom in particle accelerators, and Edward Victor Appleton for demonstrating the existence of the ionosphere. One of his most famous statements is quoted, "*In science there is only Physics; everything else is stamp collecting.*" He died in Cambridge on October 19, 1937.

We return to the discovery of the new atomic model.
Rutherford's experiment was carried out by firing alpha particles (α) at gold flakes, some tens of atoms thick.
Alpha particles (α), also called rays α, are nothing but Helium nuclei, consisting of 2 protons and 2 neutrons, thus, positively charged particles.
As a result he observed a coloumbian scattering phenomenon, that is, a deviation of the path of these particles in only 1% of cases, while for 99 percent they continued undisturbed.

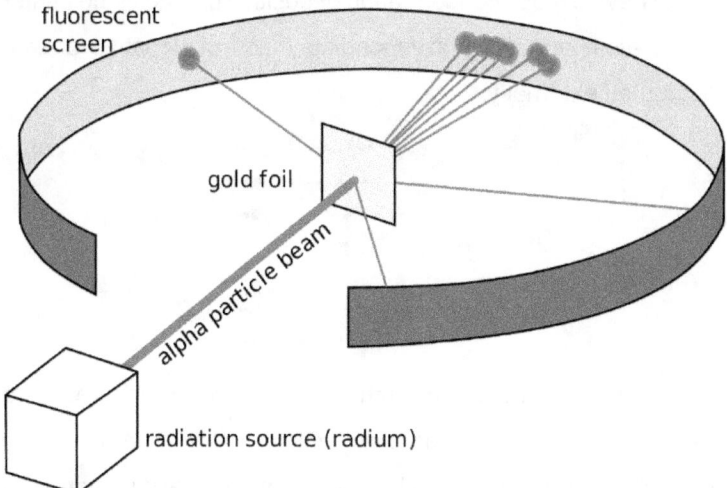

The angle of deviation ranged from 0° to 180°, in the respective limiting cases of undisturbed particle passage and return in the same direction with opposite direction.

From this he noted that the atom could not have a "pancake" configuration, with a uniformly distributed positively charged mass as Thompson thought, otherwise the particles □, having a positive charge, should have always passed through undisturbed due to the prevalence of the mass of the projectile particle (□) over the distributed mass.

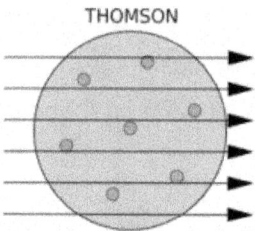

Instead, the new atom, consistent with experimental experience, had to have a mass concentration of positive charge in the center of the atom and the corresponding mass of negative charge distributed externally.

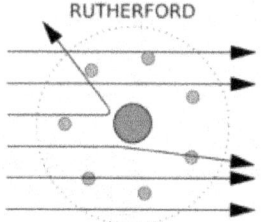

The projectile particles that struck the positively charged core would receive an appropriate deflection greater than 0° and up to 180°, in the case of returning in the same direction with opposite direction.

Conversely, particles passing through the atomic part occupied by the far less massive distributed negative charges would have passed through undisturbed.

Thus was born the Rutherford atomic model, consisting of a positively charged central nucleus in addition to negatively charged electrons orbiting around it. This atomic model, because of similarity to the motion of the planets around the sun, came to be called the planetary atomic model.

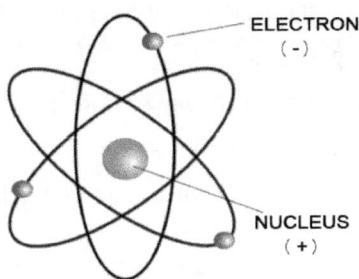

For such an atomic model, it is simple to calculate the total energy of the orbiting electron as a function of radius and electric charge.

The electron is subject to a centripetal force, materialized by the force of electrostatic attraction.

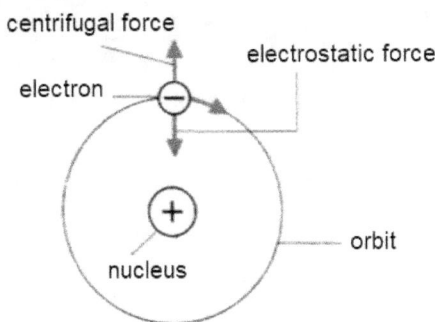

From Coulomb's relation know that nucleus and electron attract each other with a force equal to:

$$(2.1.1) \quad F = - \frac{1}{4\,\pi\varepsilon_0} \frac{e^2}{r^2}$$

with

e = charge of the electron

ε_0 = dielectric constant in vacuum

r = radius of the orbit

Since this relationship is similar to Newton's gravitational law, the orbits of electrons should logically have been elliptical in analogy to the motion of planets.

As a first approximation, however, we disregard the elliptical shape and use the circular shape.

Under such conditions the centripetal force holds:

$$(2.1.2) \quad F = -m\frac{v^2}{r}$$

with

v = tangential velocity

m = mass of the electron

Equalizing (2.1.1) with (2.1.2) gives:

$$(2.1.3) \quad \frac{1}{4\,\pi\varepsilon_0}\frac{e^2}{r^2} = m\frac{v^2}{r}$$

From which dividing by 2 and multiplying by r both members

$$(2.1.4) \quad \frac{1}{2}\,m\,v^2 = \frac{1}{8\,\pi\varepsilon_0}\frac{e^2}{r}$$

Which is precisely equivalent to kinetic energy, at non-relativistic speeds. Potential energy is worth

$$E = \frac{1}{4\,\pi\varepsilon_0}\frac{e^2}{r}$$

Ultimately, the total energy of the electron is equal to

$$E_t = \frac{1}{2}\,m\,v^2 - \frac{1}{4\,\pi\varepsilon_0}\frac{e^2}{r}$$

In the previous report we find that potential energy has a minus sign, as it relates to opposite charges of the attractor type.

Substituting (2.1.4) into the last relation found on the total energy of the electron orbiting the nucleus, we obtain

$$(2.1.5)\ E_t = -\frac{1}{8\,\pi\varepsilon_0}\frac{e^2}{r}$$

Unfortunately, this model also continued to have physical problems.

It was already known how a charged particle, when accelerated, emitted energy in the form of electromagnetic radiation, losing energy.

In the specific case, the motion being circular, generating centripetal acceleration in the motion (centrifugal from the electron's point of view), the electrons revolving around the central nucleus should have lost energy until collapsing onto the nucleus in a very short time.

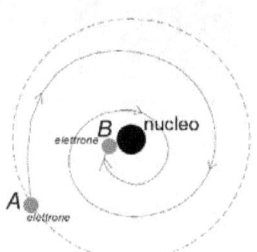

Electron in position A collapses to position B

Rutherford's model, in addition to remaining a classical model, respectful of the physical laws of classical mechanics and related to a motion assimilated to that of the planets, was largely disrespectful of experimentally obtained results.

"All science is either physics or stamp collecting"
ERNEST RUTHERFORD
https://www.goodreads.com/author/

2.2 ULTRAVIOLET CATASTROPHE

For a proper identification of an atomic model, which also had experimental verification, it was necessary to abandon certain prejudices set by classical mechanics, especially to overcome the stumbling block of the predicted energy loss of the electron to its collapse on the atomic nucleus.

A first major change came at the hands of the German physicist, Max Planck, who formulated a law of quantization of the energy of an electromagnetic wave, in the course of solving a problem in the thermodynamic field, called "ultraviolet catastrophe," related to the study of the blackbody spectrum.

Max Planck was born in Kiel on April 23, 1858, and was always considered from his high school studies to be a clear, logical and versatile mind. Appointed professor of theoretical physics at the University of Berlin, he devoted himself, mainly because he was  interested in the first incandescent lamps, to the study of thermodynamic problems related to radiation. December 14, 1900, with the publication of his first paper on quantum theory, represents the birth date of modern physics. A revolutionary in spite of himself, he was almost convinced that the concept of "quantum" was just a "fortunate purely mathematical violence against the laws of classical physics." About the theory on the interpretation of the blackbody spectrum, he wrote: "*The whole thing was an act of desperation.... I am a quiet scholar, by nature averse to rather risky adventures. But ...... a theoretical explanation had to be given, whatever the price....* In the theory of heat it seemed that the only things to be saved were the two fundamental principles (conservation of energy and the principle of entropy), for the rest I was ready to sacrifice all my previous convictions." And again in a passage from his last lecture, a few months before his death, "*Those who are engaged in the construction of the sciences will find their joy and happiness in having investigated the investigable and honored the unobservable.*" After much spiritual and material suffering, Planck spent the last years of his life in Gottingen, where he died almost 90 years old on Oct. 4, 1947.

Let us return to the issue of the blackbody spectrum.
From the application of Maxwell's equations, it appeared that a black body in thermal equilibrium with the environment, thus at a fixed temperature, the energy emitted for each radiation at different frequencies/wavelengths (radiance), should be inversely proportional to the wavelength.
Specifically, the Rayleigh-Jeans Law correlated energy density with the corresponding wavelength (λ), through the Boltzmann constant (k) and temperature (T)

$$\frac{dE}{d\lambda} = \frac{8\pi kT}{\lambda^4}$$

This relationship gave comforting results when operating with high wavelength radiation on bodies at room temperature, but instead came into crisis with the experimental results obtained from the analysis of a black body, where temperatures are significantly higher and the spectrum of radiation involved is broader.

An ideal black body is a body that does not exist in nature; it is certainly not black in color. It is an ideal body that absorbs all incident electromagnetic radiation without reflecting it.

Absorbing all the incident radiation, by the law of conservation of energy, the black body radiates the same amount of absorbed energy, although transforming it.

In the laboratory, a black body, can be made as a hollow object, insulated with the outside and kept at a constant temperature, like a kind of oven.

The object made in this way, in order to be able to proceed with the necessary experimental verifications, has a tiny hole for the entry of electromagnetic radiation, of such a size that it has minimal probability of exiting.

The inner walls absorb and emit part of the radiation, continuously, in the different wavelengths and frequencies, for any given temperature value.

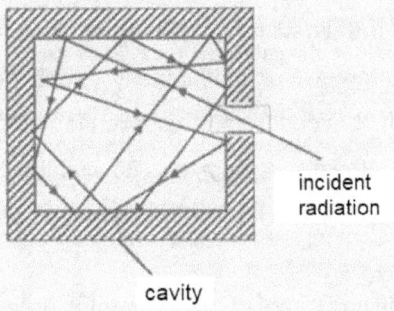

incident
radiation

cavity

The tiny hole is thus also used as a peephole, to analyze the distribution of the electromagnetic spectrum of radiation inside the hollow body.

From the above observation, a graph can be constructed that correlates the possible values of energy emitted for each radiation at different frequencies/wavelengths (radiance), as temperature changes.

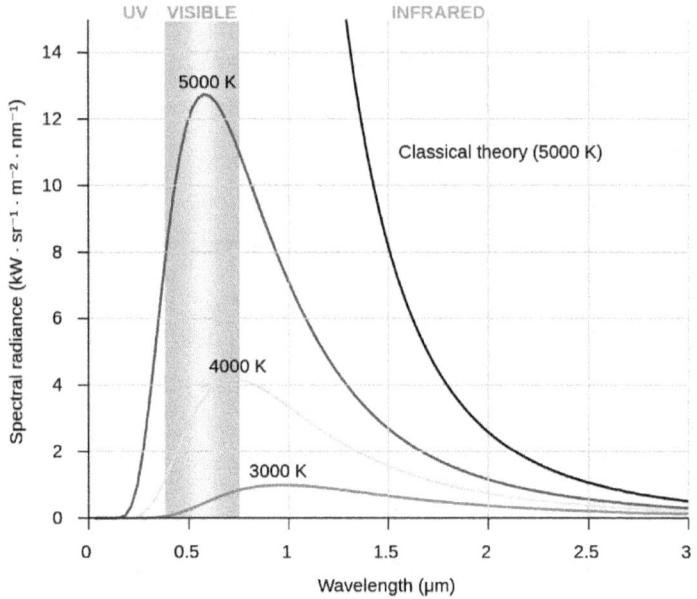

The graph shows that as the wavelength decreases, and thus moving toward the ultraviolet (UV) lengths, located to the left of the visible spectrum, we obtain intensity values that tend to zero, unlike the result obtained from the application of the old classical theory, where intensity values tend to infinity because of the inverse proportionality relationship between wavelength 'and energy.

The obvious discordance of experimental data obtained with classical theories, corresponding to radiation of wavelengths toward the ultraviolet, has led to this issue being named the "ultraviolet catastrophe."

To resolve the issue, Max Planck thus intervened, through the formulation of an innovative hypothesis for the quantization of electromagnetic radiation, which would lead, among other things, to the formulation of a new law as an expression of the energy emitted by radiation at different frequencies $B(v,T)$, as a function of temperature and frequency, using Boltzmann's constant (k), the constant speed of light in vacuum (c) and a new Planck constant (h)

$$(2.2.1)\ B(v,T) = \frac{2\,h\,v^3}{c^2}\ \frac{1}{e^{\frac{hv}{kT}} - 1}$$

And it is in the latter relationship that the constant **h**, called Planck's constant, is introduced as a pivotal element for the quantization of electromagnetic radiation, which is discussed in more detail in the next section.

Planck's constant h, represents the minimum possible action, defined as the "quantum of action," which calculated experimentally takes a constant value of 6.62606957 x 10^{-34} J s, and is the most important constant of quantum mechanics , just as the constant c, speed of light in a vacuum, represents the most important constant for the study of Einstein's Relativity.

The introduction of Planck's constant h, enshrines the actual innovation of quantum mechanics vis-à-vis classical mechanics, establishing that energy and the fundamental physical quantities related to it, with evidence only at the microscopic scale, do not evolve continuously, but are quantized, with energy, for example, being able to take on only multiple values of this constant.

The limit of interpretation of a physical phenomenon, between quantum theory and classical theory, is precisely the comparability of the value of the Action relative to the observed event with the value of Planck's constant.

Physical phenomena having a value of Action comparable to the constant h assume quantum-like behavior.

2.3 PLANCK'S CONSTANT

Let us examine in detail what Planck's constant actually represents from a physical point of view.

From classical physics we learned to use space-time graphs to represent the motions of material points.

Similarly, again from a classical point of view, it is possible to represent a motion of a particle in a velocity-space Cartesian axis system.

The following graph represents the one-dimensional motion of a material point in uniform rectilinear motion, constant velocity, moving from A to B, in a velocity-space reference system.

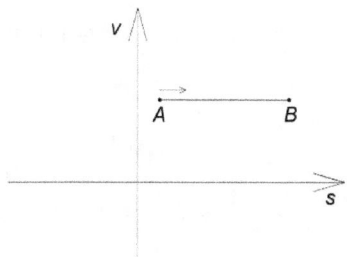

Unlike a velocity-time representation , where time is always of an increasing type, from the present to the future, in the velocity-space representation, space can take increasing and decreasing values. Such a representation system, moreover, provides information about the velocity of the particle when it is at a particular point in the considered space, unlike the velocity-time representation where it is possible to correlate velocity only as a function of time.

For simplicity, let us continue to consider a one-dimensional type of motion, of a particle moving by uniform motion in a small box with elastic bumps on the walls, such that when the particle reaches the right wall, by the law of conservation of momentum, it reverses its velocity, without varying it, and so on when it reaches the left wall.

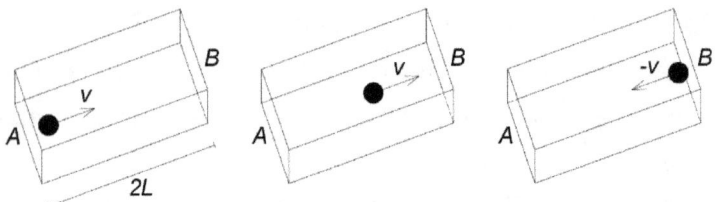

If we take the center of the box as the reference origin, assuming the box to be **2L** *in* length, the particle moves in one-dimensional space between **L** and **-L**.

The resulting motion can be represented in the velocity-space diagram as follows.

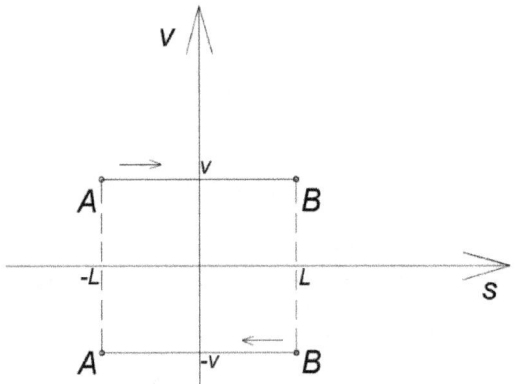

When the particle reaches the right side B, its velocity instantaneously reverses, from *v* to *-v*, and space is traveled in the opposite direction.

To take into account the mass of the particle as well, we can consider the momentum or impulse variable, equal to velocity multiplied by mass ($p=m\cdot v$), on the vertical axis in place of the velocity variable alone, without qualitative changes in the graph.

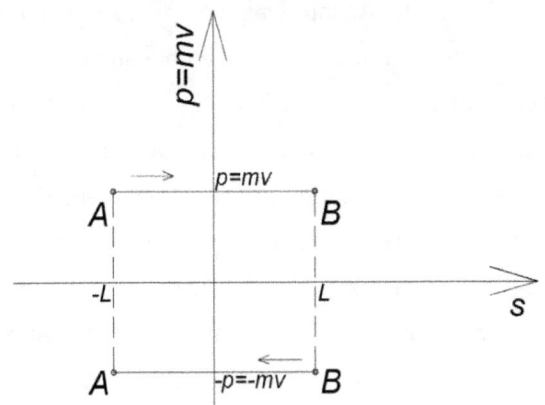

The area of the rectangle thus identified, for the one-dimensional motion considered, is called the "reduced action along a closed trajectory in phase space" and is equal to

$$(2.3.1)\ A = 2L\cdot 2\cdot m\cdot v$$

and from a dimensional point of view can be expressed as

$$[M]\ [L]\ [T]^{-1}\ [L] = [M]\ [L]^2\ [T]^{-1}$$

or rather, considering the corresponding units of measurement in the International System is expressible in

$$Kg\cdot m^2\cdot s^{-1}$$

and again, in terms of energy, taking into account that for one Joule the following conversion is valid

$$J = Kg\cdot m^2\cdot s^{-2}$$

you get

$$Kg\cdot m^2\cdot s^{-1} = J\cdot s$$

Therefore, the area identified in the "momentum-space" graph assumes the same unit as Planck's constant, which we have already called the minimum possible action and know to be approximately $h = 6.62606957 \cdot 10^{-34}$ J s.

The physical quantity "Action" was already known in the time of classical mechanics, but sparsely used because in the study of macroscopic physical laws this quantity turns out to have little use, given its small value compared to macroscopic quantities.

In fact, if we imagine that the particle is represented by a ping-pong ball and the box is a game table, assuming the weight of the ball to be *10 g*, its velocity to be *10 m/s*, and the length of the table to be *2.5 m*, applying *(2.3.1)* yields an action value of

$$A = 2L \cdot 2 \cdot m \cdot v = 2.5 \ m \cdot 2 \cdot 0.01 \ Kg \cdot 10 \, m/s = 0.5 \ Kgm^2 /s = 0.5 \ Js$$

Said value, when compared with Planck's constant *h* is about $\approx 7.50 \cdot 10^{32}$ *h* times larger.

From the value of the Action thus obtained, it is evident that in the macroscopic world it makes no sense to speak of Planck's constant since the observed phenomena present a value of the Action, abundantly multiple of said constant.

It is like looking at an expanse of sand in the desert and asking the question of how many grains of sand it consists of.

What is quite different is grabbing a fistful of sand where we realize the actual constitution in grains.

The same happens, then, when we observe the microcosm, then it makes sense to talk about the constant *h*.

If we consider an electron having mass $m = 9.11 \cdot 10^{-31}$ *Kg* moving in our hypothetical box of the size of the order of magnitude equal to 10 times the atomic radius of hydrogen $2L = 53 \cdot 10^{-11}$ m, with velocity v equal to about 1% of the speed of light c, $v \approx 3.000,000$ m/s, we obtain an action value of about $A \approx 2.90 \cdot 10^{-33}$ $J \cdot s$, evidently of the order of magnitude of the minimum action h.

In the previous discussion we have for simplicity calculated the action in one-dimensional motion, actually it is also possible to do the calculation in three-dimensional space with the help of some mathematics, but apart from the complications of calculation, in essence the concept expressed does not vary.

2.4 QUANTIZATION

Between 1900 and 1905, through the contribution of Max Planck, using the results obtained in the course of solving the problem of

 "ultraviolet catastrophe, and later with the contribution of German physicist and philosopher Albert Einstein, following his studies on the photoelectric effect, the "quantum of light" is introduced as an elementary constituent of electromagnetic radiation.

This "minimal packet or quantum" of an electromagnetic wave, endowed with both energy and momentum (mass times velocity), would only later, in about 1926, be called the "Photon." The photon concept makes electromagnetic radiation a particle, thus subjecting it to the corpuscular theory of light; such a particle is fundamentally indivisible, has zero mass and electric charge, and propagates at the speed of light.

Planck through a simple relationship correlated energy to the frequency of electromagnetic radiation by means of the proportionality constant h, appropriately named Planck's constant.

Thus the energy of an electromagnetic radiation was quantized and non-continuous, with possible values of energy proportional to a universal constant **h** and only to the frequency of the same radiation.

For one photon we can write:

$$(2.4.1) \quad E = h\,\nu$$

With

ν =   Electromagnetic radiation frequency

h =   Planck's constant

E=   Energy of a quantum of radiation e.m./photon

Planck's constant h, representing the minimum possible action, is referred to as the "quantum of action." Its constant value is calculated experimentally and is equal to 6.62606957 x 10^{-34} J s. It is reiterated how Planck's constant is the most important quantity in quantum mechanics, its introduction having sanctioned the actual innovation vis-à-vis classical mechanics. Dimensionally, this constant takes on the appearance of an energy for a time.

Also frequently appearing in treatises is the constant reduced $\hbar$, which reads h cut off and is worth $\hbar = \dfrac{h}{2\pi}$

For a number n of photons, (2.4.1) is written:

$$E = n\,h\nu$$

Frequency ν,   according to the already known knowledge of wave mechanics, can also be expressed as a function of the wavelength and speed of light in vacuum

$$(2.4.2)\ \ \nu = \frac{c}{\lambda}$$

Where the wavelength represents the distance between the two maxima or the two minima of intensity of the function describing the electromagnetic wave

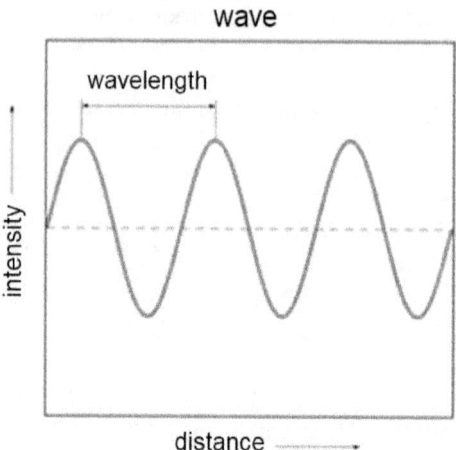

wave

wavelength

intensity

distance

(2.4.1) describes for a single photon, or rather for a single quantum of radiation, the value of quantized energy proportional to a constant *h*, the greater the frequency of the electromagnetic wave or the shorter the wavelength, without depending on the intensity of the radiation itself.

Independence from radiation intensity implies that a packet or quantum of light belonging to the visible spectrum of strong intensity possesses lower energy than a quantum of light of lower frequency, such as Laser.

In order to have energetic radiation, one must use radiation at high frequencies, not high intensities.

Looking at the electromagnetic spectrum, as the set of all possible frequencies of electromagnetic radiation, below, we find that the part on the right (X-rays, $\gamma - \rho\alpha\psi\sigma$, etc.,), where higher-frequency waves are found, identifies radiation with higher energies than the part on the left (visible light, microwaves, radio waves, etc.) having low frequencies.

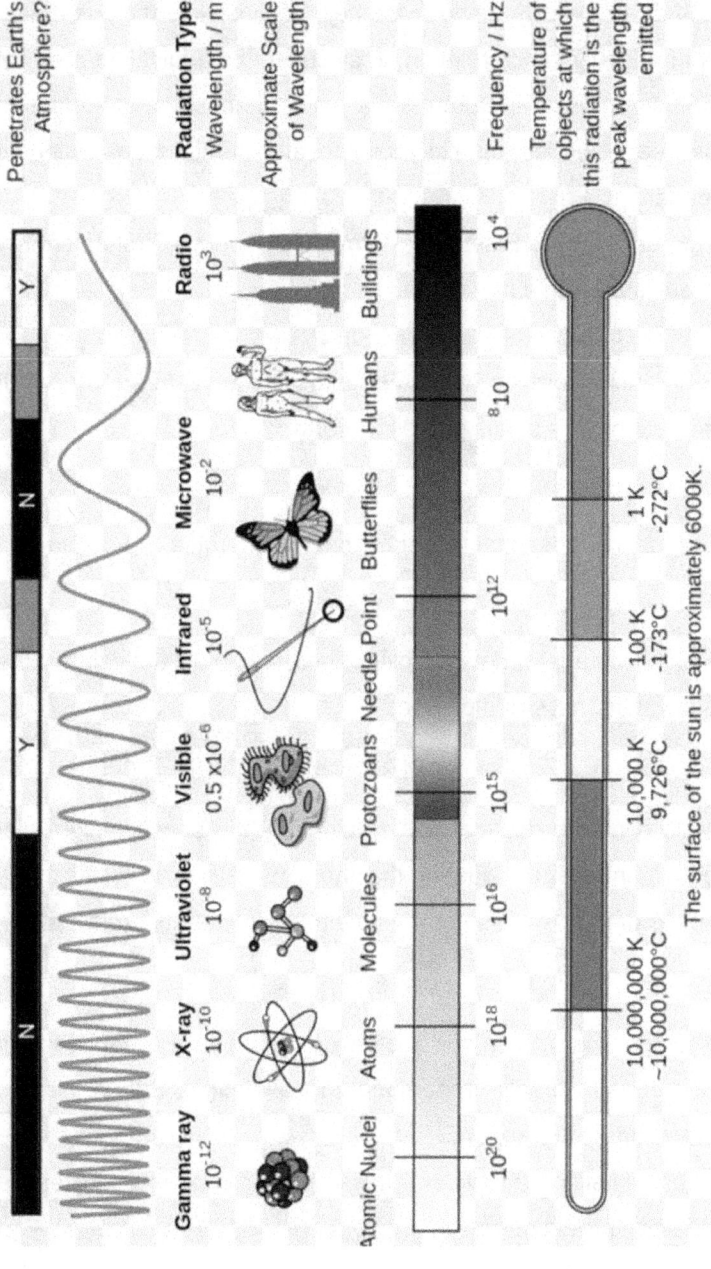

39

Penetrates Earth's Atmosphere?

Radiation Type Wavelength / m

Approximate Scale of Wavelength

Frequency / Hz

Temperature of objects at which this radiation is the peak wavelength emitted

Radiation Type	Gamma ray	X-ray	Ultraviolet	Visible	Infrared	Microwave	Radio
Wavelength / m	10^{-12}	10^{-10}	10^{-8}	0.5×10^{-6}	10^{-5}	10^{-2}	10^{3}

Approximate Scale of Wavelength: Atomic Nuclei, Atoms, Molecules, Protozoans, Needle Point, Butterflies, Humans, Buildings

Frequency / Hz: 10^{20}, 10^{18}, 10^{16}, 10^{15}, 10^{12}, 10^{8}, 10^{4}

Temperature: 10,000,000 K ~10,000,000°C; 10,000 K 9,726°C; 100 K -173°C; 1 K -272°C

The surface of the sun is approximately 6000K.

(2.4.1) also allows us, with the help of Einstein's famous mass-energy equivalence relation, to calculate the equivalent mass of a photon, which is useful for a quick dimensional comparison between electromagnetic radiation and particles consisting of matter.

We start with Einstein's formulation of the total energy associated with a moving mass

$$E = m \, c^2$$

By Planck's formula, written by substituting (2.4.2) into (2.4.1), the energy of a single photon is worth

$$E = h \, \frac{c}{\lambda}$$

Comparing these last two equations on energies, we obtain

$$m \, c^2 = h \, \frac{c}{\lambda}$$

Making the appropriate simplifications, we obtain the value of the equivalent mass of a photon:

$$(2.4.3) \quad m = \frac{h}{c\lambda}$$

Multiplying the relation thus obtained by the speed of light c at both members, considering that the e.m. wave propagates at the speed of light, we also easily obtain the momentum relation

$$(2.4.4) \quad p = m \, c = \frac{h}{\lambda}$$

The equivalent mass of a photon, as can be easily deduced from (2.4.3), is inversely proportional to wavelength, that is, directly proportional to frequency.

The greater the equivalent mass of a radiation the greater the effects it produces.

That is why in order for the photon to make its presence and effects felt, it needs to have a high frequency or low wavelength. High-frequency electromagnetic radiation is called ionizing because it succeeds in tearing electrons from the atom, due to its high energy power and equivalent mass value comparable to the mass of the electron with which it interacts.

Visible light, on the other hand, fails to make its effects felt, because its equivalent mass is about 200,000 times smaller than the already small electron, as can be easily derived from the following calculation.

To better understand the difference in numerical terms, let us try to perform some calculations.

Assume a photon in the visible range, having wavelength $\lambda = 0.5$ 10 m.$^{-6}$

Substituting the works of λ, h and c into (2.4.3), we calculate the equivalent mass of said photon

$$m_{fv} = \frac{6.62606957 \; 10^{-34} \, J \, s}{299{,}792{,}458 \frac{m}{s} \; 0.5 \; 10^{-6} \, m} = 4.42 \; 10^{-36} \frac{J \, s^2}{m^2}$$

Being $1 J = 1 \frac{Kg \, m^2}{s^2}$ we have

$$m = 4.42 \; 10^{-36} Kg$$

Knowing the mass of the electron to be 9.11 - 10⁻³¹ we can calculate the ratio (electron mass)/(visible photon equivalent mass)

$$r = \frac{m_e}{m_{fv}} = \frac{9{,}11 \; 10^{-31}}{4{,}42 \; 10^{-36}} = 206{,}108.60$$

This result evidences an equivalent mass of the photon in the visible range far less than the mass of the electron, such that its possible interaction is averted.

In contrast, if we consider high-frequency electromagnetic radiation, such as γ gamma rays, we have

$$\lambda = 10 \text{ m}^{-12}$$

$$m_{f\gamma} = \frac{6.62606957 \ 10^{-34} \text{ J s}}{299,792,458 \frac{m}{s} \ 10^{-12} \ m} = 2.21 \ 10^{-30} Kg$$

the electron mass/electron mass ratio eq. Photon γ

$$r = \frac{m_e}{m_{f\gamma}} = \frac{9.11 \ 10^{-31}}{2.21 \ 10^{-30}} = 0.41$$

The result obtained evidences an electron mass value lower than the equivalent mass of the γ photon, consistent with the ionizing properties of γ rays.

The possible interaction of electromagnetic waves with matter places constraints on measurement processes in the microscopic world.

In fact, in order to make measurements of particles, such as electrons, it is necessary to use high-frequency, low-wavelength electromagnetic waves so that their position, for example, can be detected.

Conversely, the use of such type of radiation, with high power of interference with the very motion of the particle, leads to results influenced by the measurement process, unlike in classical physics where the radiation used for measurement processes is at high wavelength (visible light) with no power of interference towards the observed physical state.

Performing the equivalent mass calculation for *X-rays*, which have an average wavelength value of $\lambda = 10^{-10}$ m, we obtain an electron to X-photon equivalent mass ratio of 41.

Said value manifests a lower interaction value of *X* photons with matter, compared with γ rays, and thus a lower ionizing power. This peculiarity of X-rays is exploited in the medical field for taking X-rays of biological bodies.

The operation is based on the interaction between a beam of energetic photons, precisely X-rays, directed from a source to a receptor, with the interposed matter (biological body).

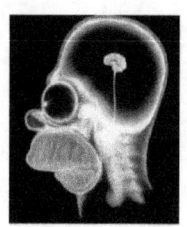

The atoms of such an interfering body, exclusively in areas of high atomic density, prevent the photons from reaching the receptor, resulting in a faithful image of the biological body "in negative," only the photons that are not absorbed instead being imprinted on the film.

This practice provides only morphological information about the biological body, such as the presence of bone fractures or thickened masses.

The amount of radiation is well dosed and in limited amounts, such that in terms of comparison, an intercontinental, round-trip air flight from Europe to America is equivalent to taking 5 chest X-rays.

CT scan, which stands for COMPUTERIZED AXIAL TOMOGRAPHY, also uses X-rays, in a more advanced way.

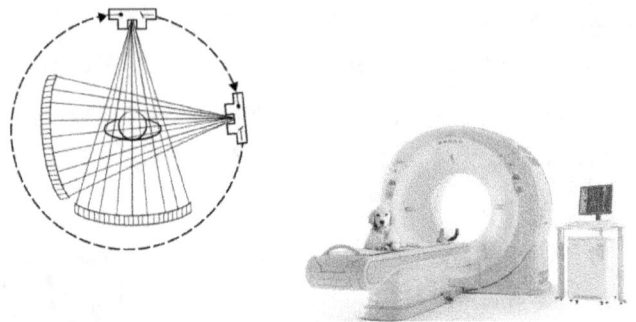

Through the use of a moving source, multiple body sections or layers (tomography) of the patient can be reproduced and consequently three-dimensional processing can be carried out. Undergoing a CT scan deserves a little more attention considering that a CT scan of the chest is equivalent to taking about 385 chest X-rays.

It is said that a visitor once came to the home of Nobel Prize–winning physicist Niels Bohr and, having noticed a horseshoe hung above the entrance, asked incredulously if the professor believed horseshoes brought good luck. "No," Bohr replied, "but I am told that they bring luck even to those who do not believe in them"

NIELS BOHR

https://www.laphamsquarterly.org/magic-shows/miscellany/niels-bohrs-lucky-horseshoe

2.5 BOHR'S ATOMIC MODEL

The Danish physicist, Niels Bohr, in 1913 solved the inherent problem of the electron falling on the nucleus and of some experimental discrepancies on emission spectra by proposing appropriate variations from the previous atomic model.

Bohr was born in Copenhagen on October 7, 1885. His father Christian Bohr was a Danish physiologist, professor of physiology at the University of Copenhagen and discoverer of a behavior of hemoglobin called the Bohr effect. His paternal grandfather Henrik Bohr was a  teacher and later dean of the Westenske Institut in Copenhagen. His mother, Ellen Adler Bohr, was a wealthy Danish bourgeois of Jewish origin whose family was very prominent in Danish banking and parliamentary circles.

His brother, Harald Bohr, was a mathematician and soccer player for the Danish national team who was summoned to the Olympics. Niels was a soccer player like his brother, but an amateur, playing goalkeeper, and played in 1905 with his brother on one of the Copenhagen teams.

Bohr graduated from the University of Copenhagen in 1911. He moved first to Cambridge on a fellowship, where he hoped to collaborate with J. J. Thomson to continue investigations into the theory of metals. Failing to work with the British physicist, he ventured into the study of electromagnetism. Thanks to another fellowship, he then moved to the University of Manchester, England, where he studied with Ernest Rutherford. During his time studying with Rutherford, he was involved in the successful completion of some experiments on the absorption by aluminum of alpha particles, a program suggested by Rutherford himself. This project was later suspended by Bohr himself because he was interested in the theoretical concept of his new atomic model, which originated from the orbital theory of the atom discovered by Rutherford. Many years after Rutherford's death, Bohr agreed to deliver his memorial address, known as the Rutherford Memorial Lecture, on November 28, 1958, at Imperial College, London. Albert Einstein was also a friend of Bohr's, and it was in a letter to him in 1926 that Einstein made his famous remark about quantum mechanics, often paraphrased as "God does not play dice with the universe," to which he replied, "Don't tell God how he should play." He died in Copenhagen on November 18, 1962.

Let us return to Bohr's proposed solution for the description of the new atomic model.
This, consists of the proposal of a model with quantized energy and orbits, following in the wake of the results obtained, on the subject of quantization by Max Planck from Albert Einstein.
Under such a hypothesis, electrons are only allowed to occupy multiple spatial sectors of discrete values, so as to force the electron not to flow into the central nucleus.
Thus continues the replacement of the concept of continuity in favor of a process of discretization of natural phenomena.
The electron, now occupies a quantized orbit with a well-established energy value, which is also quantized.

The same electron can change orbit, but it will be necessary to supply or subtract energy.

In an analogy in the classical world, we can think of discrete values of energy in a stepped form, unlike the continuous values of energy that can be depicted with an inclined line.

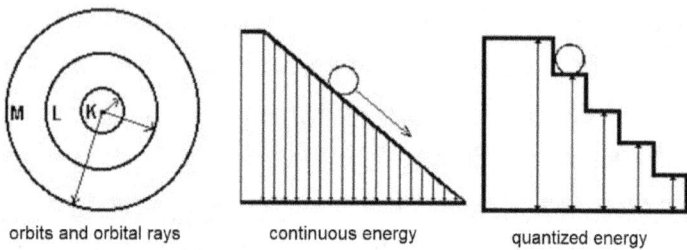

orbits and orbital rays   continuous energy   quantized energy

A ball placed on steps gains speed as it descends, gaining kinetic energy. In attempting to make the ball rise, however, it will be necessary to administer kinetic energy.

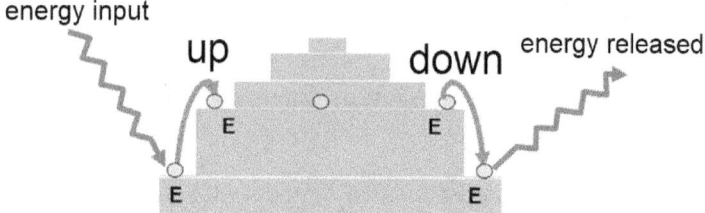

In the same way as the ball in the previous example, the electron can change orbit, only if there is a quantized energy transfer or input with or to the outside world, of an amount at least equal to the height of the corresponding energy step.

If the electron goes from a higher level where it is, to a lower one, it will have to lose energy, materialized through the emission of a *"flash of light"* (photon).

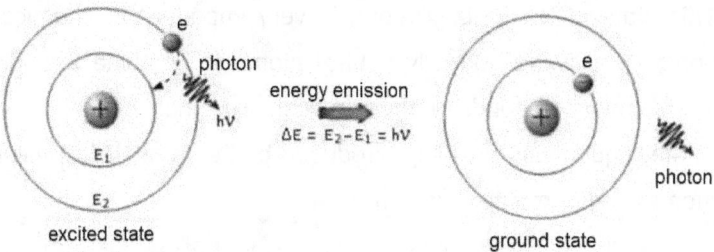

Otherwise, by inputting energy, either through the input of a photon or through particle-to-particle collisions, the electron makes a quantum leap from a lower to a higher energy level, and the atom is called "excited."

This excited state, however, is unstable, with the consequence that the electron tends to return to its initial position, returning the acquired energy, through the emission of the previously acquired photon.

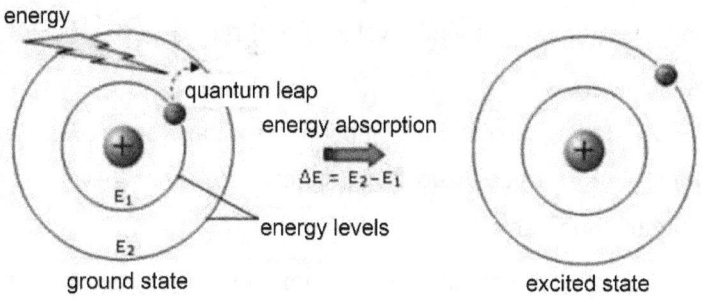

If the energy input is excessively high, it is possible for the electron to be torn from the orbital and the atom remains positively charged, having lost the negative electron. In the latter case, the atom is said to become ionized.

This state of the ionized atom, is very important in chemical bonding, allowing normally neutral atoms to combine through electromagnetic bonds for the composition of molecules.

Another quantized variable introduced by that model is angular momentum or momentum of motion.

$$(2.5.1) \quad \vec{L} = \vec{r} \times \vec{p} = \vec{r} \times m\,\vec{v}$$

This variable being a vector, has the classical three components along the three Cartesian axes x, y and z.

Recall that a vector is a geometric entity characterized by direction, direction and intensity, suitable for representing physical quantities in space.

Considering the modulus of angular momentum, with respect to the center of rotation, Bohr, formulated its quantized scalar value

$$(2.5.2) \quad L = m\,v\,r = n\,\hbar$$

With $n = 1,2,3 \dots$ (integers), $\hbar$ is Planck's reduced constant, m, v and r are mass, tangential velocity and radius of the electron, respectively.

A close deduction of this assumption is that the orbital radius also turns out to be quantized and a function solely of the principal quantum number n.

By quantizing the angular momentum and radii of the orbits, the problem of the expected fall of the electron on the nucleus can be solved.

From an analytical point of view, for the model of the hydrogen atom, setting as the equilibrium position the equality of the centripetal force with the force of attraction of electric charges, in analogy to what has already been performed with the Rutherford model with *(2.1.3)*, and multiplying at both members by r^2 , we obtain

$$(2.5.3) \quad \frac{1}{4\,\pi\varepsilon_0}\, e^2 = m\, v^2 r$$

Substituting *(2.5.2)* into the previous *(2.5.3)*

$$\frac{1}{4\,\pi\varepsilon_0}\, e^2 = v\, n\, \hbar$$

Isolating the variable speed

$$v = \frac{1}{4\,\pi\varepsilon_0 n\, \hbar}\, e^2$$

Substituting this into *(2.5.3)* gives

$$\frac{1}{4\,\pi\varepsilon_0}\, e^2 = m\, \left(\frac{1}{4\,\pi\varepsilon_0 n\, \hbar}\, e^2\right)^2 r$$

$$\frac{1}{4\,\pi\varepsilon_0}\, e^2 = m\, \frac{1}{16\,\pi^2\varepsilon_0{}^2 n^2 \hbar^2}\, e^4 r$$

Simplifying

$$1 = m\, \frac{1}{4\,\pi\varepsilon_0 n^2 \hbar^2}\, e^2 r$$

Isolating the variable radius r and substituting for h the value $\hbar = \frac{h}{2\pi}$

$$(2.5.4) \quad r = \frac{\varepsilon_0 h^2}{m\, e^2}\, n^2 = k\, n^2$$

We have thus obtained, a relationship where the atomic radius depends solely on the quantized variable n =1,2,3, etc., while the other values are all constants.

Substituting *n=1* gives the measure of the minimum distance of the electron from the nucleus, in the hydrogen atom, which is called the Bohr radius.

This calculated value turns out to be in perfect agreement with the experimental data.

Ultimately in Bohr's atomic model, assigned the principal quantum number n, the radius of the orbit and the corresponding energy level are uniquely determined.

Using the quantized orbital radius relation in (2.5.4), the value of the corresponding quantized energy can be easily calculated.

Substituting (2.5.4) into the relation finding the total energy, in the case of the hydrogen atom, expressed by (2.1.5) gives:

$$(2.5.5) \quad E_t = -\frac{1}{8\,\pi\varepsilon_0}\frac{e^2}{k\,n^2} = k'\frac{1}{n^2}$$

The value of the total energy in a quantized orbital can be expressed through an inverse proportionality relationship to the square of its quantum number.

From the latter, it is possible to calculate the necessary frequency (ν) or wavelength (λ) of radiation, of a single photon, to be used to make an electron perform a quantum leap, that is, as they say in the jargon, "excite" it.

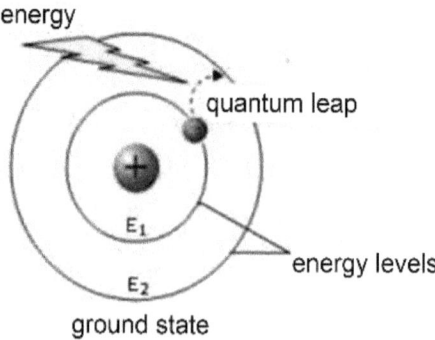

For this purpose, it will first be necessary to calculate the change in energy in the quantum jump case as a result of electron excitation, applying to the final and initial states the result obtained from Bohr's quantization assumptions, given in relation (2.5.5),

$$(2.5.6)\ \Delta E_t = k'\frac{1}{n_f^2} - k'\frac{1}{n_i^2} = k'\left(\frac{1}{n_f^2} - \frac{1}{n_i^2}\right)$$

With n_f and n_i, the initial and final energy levels, respectively. Expressing the value of energy, yielded or absorbed, as a function of frequency, through Planck's law

$$\Delta E = h\nu$$

and making it explicit in the frequency variable ν

$$(2.5.7)\ \nu = \frac{\Delta E}{h}$$

By substituting (2.5.6) into (2.5.7), we obtain

$$(2.5.8)\ \nu = \frac{k'}{h}\left(\frac{1}{n_f^2} - \frac{1}{n_i^2}\right)$$

By amalgamating all the constant values into a new constant called R, we thus obtained the sought-after relationship expressing the photon frequency required to cause a quantum jump from orbital n_i to orbital n_f.

$$\nu = R\left(\frac{1}{n_f^2} - \frac{1}{n_i^2}\right)$$

The same relationship can be expressed in terms of wavelength by substituting the known frequency/wavelength ratio from (2.4.2) into (2.5.8)

$$\frac{c}{\lambda} = \frac{k'}{h}\left(\frac{1}{n_f^2} - \frac{1}{n_i^2}\right)$$

By merging all the constants into a new constant named R', we have

$$(2.5.9) \quad \frac{1}{\lambda} = R'\left(\frac{1}{n_f^2} - \frac{1}{n_i^2}\right)$$

The latter is in harmony with the paper by Swedish physicist and mathematician Johannes Robert Rydberg, also known as Janne, who formulated his paper for the description of the spectrum of the hydrogen atom-that is, all possible wavelengths of light that the hydrogen atom is capable of emitting.

Ultimately, Bohr's atomic model starting from postulates of quantization of energy and orbital radius, as a function of the principal quantum number n, succeeds well in describing the behavior of the hydrogen atom, or for that matter any other type of atom having only one orbiting electron (hydrogen atoms).

For multi-electron atoms, on the other hand, this model failed to give comforting results compared with experimental ones, so it necessarily needed to be refined.

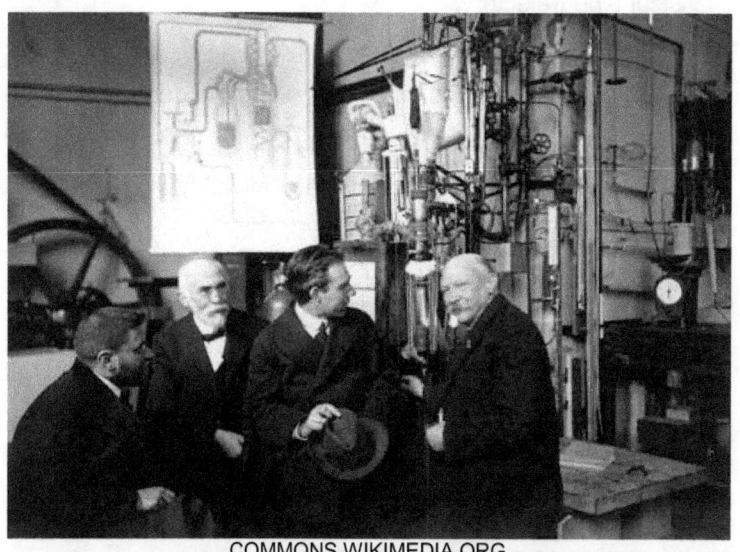

COMMONS.WIKIMEDIA.ORG
"An expert is a person who has made all the mistakes that can be made in a very narrow field"

NIELS BOHR

https://www.goodreads.com/author/

2.6 QUANTUM ATOMIC MODEL

While Bohr's atomic model included the innovative quantization of energy and angular momentum, it still remained anchored to the classical idea of the electron in orbit, following a definite classical-type trajectory.

The quantization assumptions introduced in Bohr's atom, however, formed the basis for a more elaborate quantum model, also called the "model according to the interpretation or Copenhagen school," in honor of the capital city that gave birth to Niels Bohr.

Through the contribution of further theories formulated by other distinguished scientists such as Pauli, Dirac, Sommerfeld, Heisenberg, and Schrödinger, a new "quantum" atomic model called the "standard model (MS)" or as already mentioned "model according to the Copenhagen interpretation or school," which is considered one of the most popular models underlying quantum physics studies, is thus defined.

The new atomic model becomes more complex and suitable for describing the behavior of even multi-electron atoms, with positive experimental results of atomic emission spectra.

In the quantum atom, the electron no longer has a specific trajectory, but occupies certain areas called orbitals, in the formation of the atom.

The new quantum formulation of the atom is based on the hypothesis of quantization of additional descriptive elements of atomic structure, compared to the simpler Bohr atom.

It is 'possible to define appropriate state quantities, also quantized, called quantum numbers, much used particularly in chemistry, denoted by the letters: n, l, m, s.

Quantum numbers become representative of the quantization of the energy, shape and orientation of orbitals, as well as the quantization of the intrinsic angular momentum called Spin.

The principal quantum number, denoted by the letter n, represents the quantized energy level as already used in Bohr's atomic model in *(2.5.5)*.

Quantization of orbitals in shape and orientation, results in the identification of a particular orbital space, suitably conformed and spatially oriented, occupied by electrons for any given energy level.

Associated with the shape is the secondary quantum number denoted by the letter l, which is representative of the angular momentum or momentum momentum. The latter describes a kind of distortion of the shape of the orbital with respect to spherical symmetry.

The use of the secondary quantum number, derives from similar considerations introduced by the German physicist Arnold Johannes Wilhelm Sommerfeld (during his tenure as a theoretical physicist at Ludwig Maximilian University of Munich he had among his students Werner Heisenberg and Wolfgang Pauli, whose doctoral dissertations he supervised).

Sommerfeld hypothesized that electrons traveled around the nucleus in elliptical orbits, in analogy to planetary orbits, rather than in circular orbits.

Said elliptical orbits could have different ratios of the semi-axes, thus resulting in different flattening.

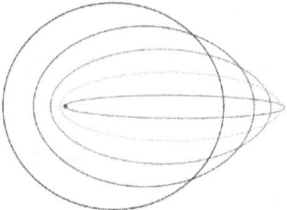

Similarly to Sommerfeld's hypothesis, the secondary quantum number l is introduced for the quantum atomic model, indicative of the amount of crushing with respect to the undeformed spherical shape.

Its value can vary as an integer depending on the main quantum number: 0 to $n-1$

$$(2.6.1) \ 0 \leq l \leq n-1$$

On the basis of this relationship, it is obtained that for the first energy level $n=1$, the secondary quantum number can take only null value $l=n-1=1=0$. The null value of l corresponds precisely to an absence of spatial distortion of spherical symmetry.

Consequently, l comes into play from the second energy level onward, where since the value of the main quantum number n is equal to 2, the quantum number l for $(2.6.1)$ can take values of 0 or equal to 1.

Denoted by the letter s, anticipated by the value of the principal quantum number n, the atomic configurations characterized by a value of the secondary quantum number null, $l=0$, are at complete spherical symmetry of the orbital shape.

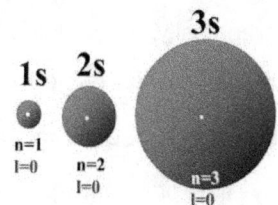

Orbitals having $l=0$ are then referred to as 1s, 2s, 3s ...etc, as the energy levels vary, that is, as the principal quantum number varies.

The value $l=1$, indicates orbitals with a partcicular distortion with axial symmetry. These orbitals are represented by the letter p, anticipated by the principal quantum number and followed by the axis of reference symmetry, such that they are referred to as 2p 2p $2p_{xyz}$, 3p 3p $3p_{xyz}$, ...etc.

The p-type orbitals are divided into three sublevels each symmetrical with respect to a Cartesian axis.

Each of the three p-type orbitals has a nodal plane: the plane passing through the nucleus and perpendicular to the symmetry axis of the orbital, understood as the geometric place where the electron has no chance to occupy its space.

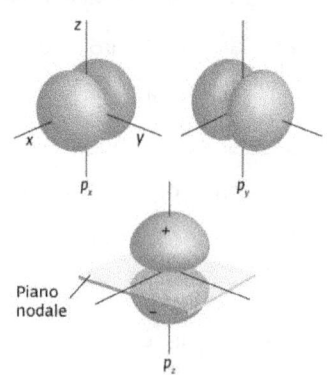

The three p orbitals possess the same energy, and since orbitals having the same energy are called degenerate, we will say that the p orbitals are three times degenerate.

Later values of l = 2, 3, 4, represent more articulated forms, such that for values of 3 and 4 the orbitals cannot be represented graphically because they are very complex.

Specifically for l=2, each orbital has two nodal planes or a nodal surface. These orbitals are named with the letter d, anticipated by the principal quantum number and followed by letters identifying the axes of symmetry: $3d\ 3d\ 3d_z{}^2{}_{xz\ yz}\ 3d_{xy}\ 3d$, $4d\ 4d$ $4d\ 4d_x{}^2{}_{-y}{}^2{}_z{}^2{}_{xz\ yz\ xy}\ x^2{}_{-y}{}^2$ $4d..etc.$

The *d-type* orbitals are five times degenerate.

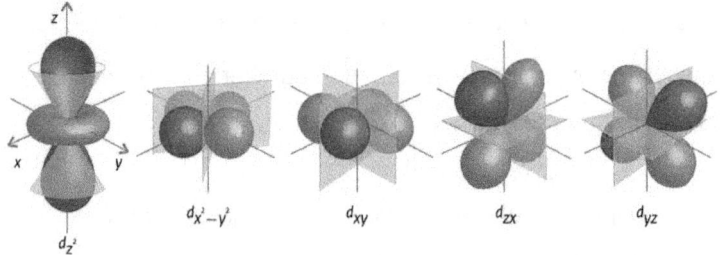

Orbitals with secondary quantum number l=3, identified with the letter f, are considered quite rarely in theoretical chemistry.

Orbitals with l=4, denoted by the letter g, are generally ignored altogether even though theoretically possible.

The orientation of the orbital, which makes it possible to distinguish the already seen sublevels for each type of orbital, is represented by the magnetic quantum number "m."

The magnetic quantum number **m** can vary as an integer between -l and +l:

$$(2.6.2)\ -l \leq m \leq +l$$

The value of the magnetic quantum number m identifies the orientation of the orbital, through the identifying value of the axis of symmetry, placed as subscript of the corresponding letter of the orbital: $2p\ 2p\ 2p_{xyz}$, $3d_z{}^2\ 3d_{xz}\ 3d_{yz}\ 3d_{xy}\ 3d_{x^2-y^2}$ ...etc.

Let us proceed to a summary by means of an explanatory example. at a principal quantum number $n=1$, for (2.6.1) and (2.6.2) it happens that the only possible secondary quantum number is equal to $l=0$, as well as $m=0$, consequently the orbital assumes the shape with spherical symmetry.

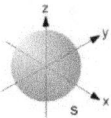

At a state with a principal quantum number of $n=2$, by (2.6.1) we can have that l can take on the values 0 and 1 only.

Consequently, for the same state, m at the value $l=0$, by (2.6.2), takes on a null value and at the value $l=1$, it can take on respective values -1, 0 and $+1$.

For $l=0$ the orbital is spherical, while for $l=1$ the orbital has a shape with axial symmetry oriented with respect to the three Cartesian axes, as the three possible values of its magnetic quantum number m=-1, m=0 and m=+1 vary.

Orbital naming includes identification of the Cartesian axis (x,y,z), placed at the subscript of the letter representing the secondary quantum state (p_x, p_y, p)$_z$.

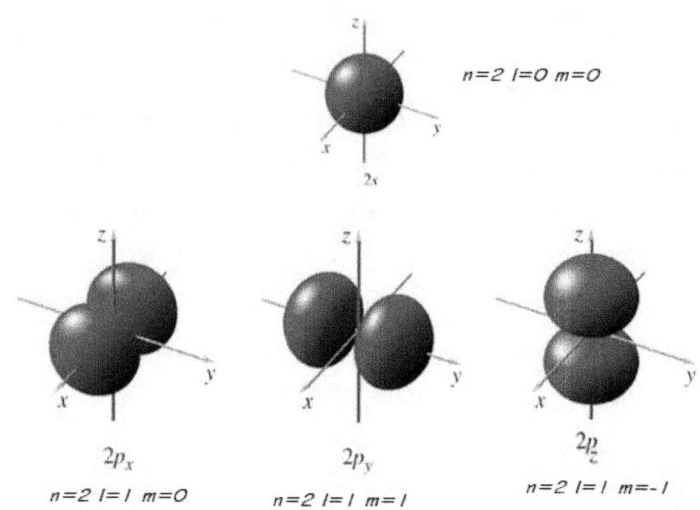

$2p_x$

$n=2\ l=1\ m=0$

$2p_y$

$n=2\ l=1\ m=1$

$2p_z$

$n=2\ l=1\ m=-1$

For a state with quantum number equal to $n=3$, we can have that l can take the values up to $n-1$, thus, by (2.6.1), the values 0, 1 and 2.

At l =0, for (2.6.2), we have that m takes on a null value. For l =1 , m can take on the respective three values -1, 0 and +1 , which correspond to p-type orbitals three times degenerate.

At l =2, again by (2.6.2), m can take as many as five values between -l to l.

The possible values of m will be -2,-1,0,1,2, corresponding to the five-fold degenerate d-type orbitals.

The following figure summarizes all the shapes and orientation of the possible orbitals, corresponding to each energy state.

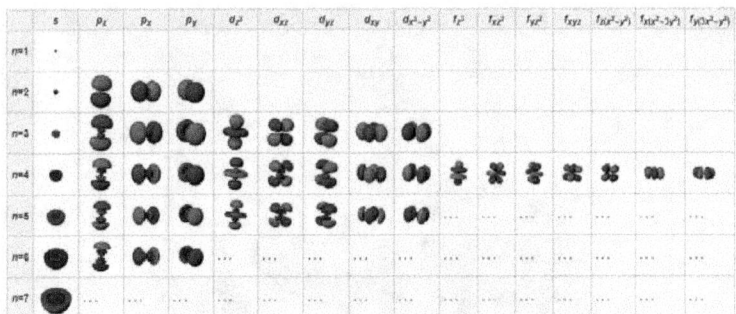

"In science one tries to tell people, in such a way as to be understood by everyone, something that no one ever knew before. But in poetry, it's the exact opposite."

PAUL ADRIEN MAURICE DIRAC
https://www.goodreads.com/author/

2.7 SPIN QUANTUM NUMBER

In the previous paragraph, the characteristics of the first three quantum numbers were set forth, leaving out the quantum number called *spin*, as an additional quantum degree of freedom of the particle.

Spin, given its greater complexity, deserves more in-depth study to understand its nature.

The "*spin quantum number,*" is denoted by the letter s, and is a number that quantizes the corresponding vector quantum quantity of a particular type. This quantum state quantity is called the "*intrinsic or spin angular momentum*" and is denoted as $\vec{S}$.

The *spin quantum number s*, on the other hand, is a scalar associated with the modulus of the spin angular momentum vector quantity.

When we talk about SPIN generically, strictly speaking we should specify whether we are talking about *spin quantum number* or *spin angular momentum.*

Simply on the basis of the value taken by the quantum number *s* it is possible to distinguish the type of particle, in a mass-independent manner.

Integer *spin* values (0,1,2,...) identify boson-type particles, while half-integer spin values (1/2, 3/2, 5/2,..) identify fermion-type particles.

The nature of bosonic and fermionic particles will be appropriately explored in later chapters.

Spin angular momentum is a physical quantity whose modulus can be expressed as a function of the corresponding *spin* quantum number and the reduced Planck constant, according to the following relationship

$$(2.7.1) \quad S = \sqrt{s(s+1)}\,\hbar$$

Unlike other quantum numbers, spin exists even for particles having zero mass.

For the photon, for example, the spin quantum number, can only take an integer value of s=1, from which applying (2.7.1) gives a spin angular momentum modulus of $S = \sqrt{2}\,\hbar$.

For the electron, on the other hand, a value of the spin quantum number of ½ is found experimentally, consequently the modulus of the spin angular momentum, applying (2.7.1) is worth

$$(2.7.2)\, S = \sqrt{\frac{1}{2}(\frac{1}{2}+1)}\,\hbar = \sqrt{\frac{3}{4}}\,\hbar = \frac{\sqrt{3}}{2}\,\hbar$$

Spin angular momentum is a particular form of angular momentum called intrinsic angular momentum, not to be confused with the angular momentum of the electron rotating around the nucleus.

The existence of a SPIN associated with particles was initially deduced theoretically and only later found experimentally, out of a need to compensate for a deficiency of only the classical angular momentum or momentum of motion.

Relative to the motion of the electron in the hydrogen atom, it was noted that the orbital angular momentum alone was not a constant of motion, as it should be.

The orbital angular momentum must be respectful of the conservation principle, the electron being in motion under conditions of a central force field, such as the proton-electron electrostatic force of attraction.

We represent on a system of orthogonal Cartesian axes, an orbiting electron, in the hydrogen atom, having velocity v however oriented, and mass m

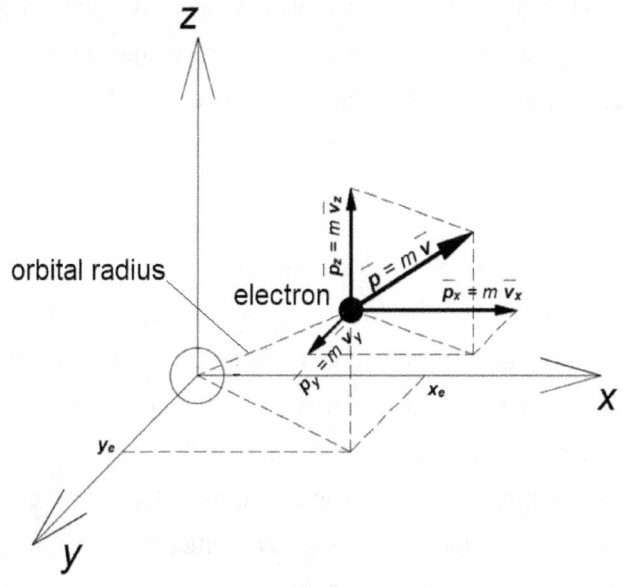

Let us examine only the component along the z axis of angular momentum.

From the above figure, knowing the direction of the moment along the z-axis, its scalar value can be derived, taking into account the perpendicularity conditions of the components

$$L_z = p_x y_e - p_y x_e = m\left(v_x y_e - v_y x_e\right)$$

Its time derivative turns out to be equal to

$$\frac{dL_z}{dt} = \frac{d\,m\left(v_x y_e - v_y x_e\right)}{dt} = m\,\frac{d\left(v_x y_e - v_y x_e\right)}{dt} = m\left(\frac{d(v_x y_e)}{dt} - \frac{d(v_y x_e)}{dt}\right) =$$

$$= m\left(v_x \frac{dy_e}{dt} + y_e \frac{dv_x}{dt} - v_y \frac{dx_e}{dt} - x_e \frac{dv_y}{dt}\right) = m\left(v_x v_y + y_e \frac{dv_x}{dt} - v_y v_x - x_e \frac{dv_y}{dt}\right) =$$

$$= m\left(y_e \frac{dv_x}{dt} - x_e \frac{dv_y}{dt}\right) = m\left(y_e \frac{d^2 x_e}{dt^2} - x_e \frac{d^2 y_e}{dt^2}\right)$$

It is evident how the latter value is zero only under certain conditions, that is, when the following equation is satisfied

$$y_e \frac{d^2 x_e}{dt^2} = x_e \frac{d^2 y_e}{dt^2}$$

For all values that do not satisfy the previous differential equation, the time derivative of the examined component of angular momentum remains non-zero, such that it can be stated that

$$\frac{dL_z}{dt} \neq 0$$

Generalizing, for all three components, it can be admitted that the orbital angular momentum alone is not conserved.

Based on these observations, it became necessary to introduce a new term so that the overall angular momentum could be a constant of motion.

So it was, that in addition to angular momentum a new entity was introduced: intrinsic or spin angular momentum.

Said intrinsic angular momentum was given the appellation SPIN, from the English "whirling spin," precisely because the particle became associated with a kind of rotation around its own axis, similar to the Earth's rotation.

Ultimately, the total angular momentum, for the orbiting electron, consists of two values, vectorially equal to

$$\vec{M} = \vec{L} \pm \vec{S}$$

Having denoted by the vector $\vec{L}$ the rate relative to the orbital angular momentum equal to $\vec{L} = \vec{r} \times m\vec{v}$ and with the vector $\vec{S}$ the intrinsic or spin angular momentum.

The carrier $\vec{S}$, unlike the vector $\vec{L}$, is a particular vector with complex components, represented, as we will elaborate later, by a linear operator of complex type operating on a state vector of the system.

The modulus of the spin angular momentum alone, on the other hand, by *(2.7.1)*, turns out to assume a real value, already expressed as a function of the real value associated with the spin quantum number.

The spin vector $\vec{S}$ is an algebraic rather than a geometric entity, so much so that it can be represented with appropriate imaginary component matrix-vectors.

While in classical physics the variables describing a state of a system are always measurable, in quantum physics a distinction must be made about the type of variable being considered.

A quantum quantity that is in some way measurable directly through appropriate measuring instruments or indirectly through analytical calculation is called an "observable."

In the case of spin, the associated quantum number takes the form of a measurable quantity and thus an "observable."

In particular, for the *spin angular momentum* vector quantity, measurements of the individual components along the Cartesian axes can be made.

In this regard we introduce the additional quantity called m_s *"magnetic spin quantum number or secondary spin quantum number,"* which, as an observable, is the value that quantizes the component along the considered axis, of the intrinsic angular momentum through the following relationship

$$S_z = m_s \, \hbar$$

The *secondary spin quantum number* m_s can take only values, integer or fractional, within the limits of the value taken by *the spin quantum number* and between *-s, (-s+1), ...(s-1),s*:

$$-s \leq m_s \leq +s$$

In the case of the electron, we have *s=1/2* and the corresponding possible values of m_s are *-1/2* and *+1/2*; consequently, the values of the two components of the *intrinsic angular momentum* along that axis, become respectively equal to $S_z = +\frac{1}{2}\hbar$ e $S_z = -\frac{1}{2}\hbar$.

In the case *s=1* we have for m_s the possible values *-1, 0, +1*.

The secondary spin quantum number m_s is of great importance, as it enables us to distinguish quantum states having the same quantum numbers.

To summarize, we have seen that the state of *spin angular momentum* is representable by a particular vector with complex components, for which only its modulus can be known.

Instead, the spin quantum *number* that quantizes the spin angular momentum is a scalar and is denoted by the letter s.

We also introduced an additional *secondary spin quantum number* denoted by the letter m_s that quantizes the component of spin angular momentum along a single Cartesian axis.

The latter is precisely the magnitude that is normally simply referred to as *spin*.

Spin is experimentally measurable as a component along a reference axis.

Let's see how the *spin* measurement mechanism works.

Experimental measurement of the *intrinsic angular momentum* of a particle, as a component along one direction, is possible because of the characteristic that the latter is a function of the magnetic moment.

In analogy to a microscopic coil traversed by current, the electron having a negative electric charge, for example, in the course of rotation about its axis, generates a magnetic dipole, that is, a magnetic field with opposite North and South poles.

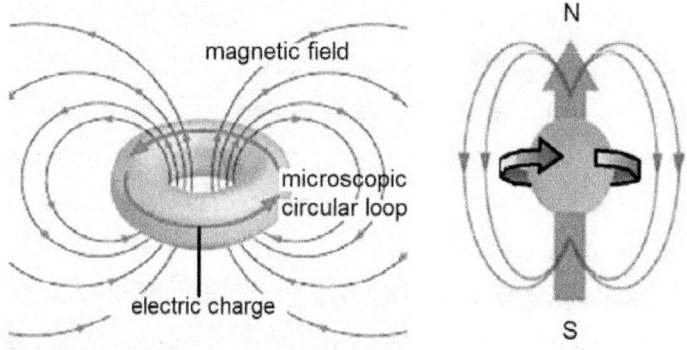

The corresponding magnetic moment is identified with a vector oriented orthogonally to the plane of rotation of the electric charge

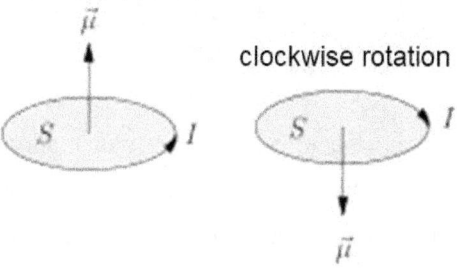

The magnetic dipole has the characteristic of orienting itself in a certain direction when immersed in a magnetic field, and it is this characteristic that allows Spin measurements to be made.

The limitation of knowing Spin's magnetic moment is that it can only be measured along one component at a time.

In fact, since the magnetic field is a vector field, applying two magnetic fields for the purpose of measurement, they would add vectorially to each other, giving rise to a new magnetic field oriented along a new resultant direction.

The measurement of the spin state of an electron, as a component along an axis, can be verified through the Stern-Gerlach experiment, already devised in 1922, by German physicists Otto Stern and Walther Gerlach.

The experimental apparatus consists of a non-uniform magnetic field generator, through which silver atoms are passed and detected on a special screen.

The measured spin value, will be relative to the unpaired electron present in the last orbital of the silver atom, the 5s orbital[1] , where only one electron is present, as can be seen from reading the periodic table.

For a particle moving in a homogeneous magnetic field, the forces exerted on the opposite ends of the dipole, north and south, cancel each other out and the particle's trajectory is not changed, which is why a nonuniform magnetic field is used for the experimental apparatus.

A particle passing through an inhomogeneous magnetic field will be subject to a force that at one end of the dipole will be slightly greater than that at the opposite end; this causes the particle to deflect.

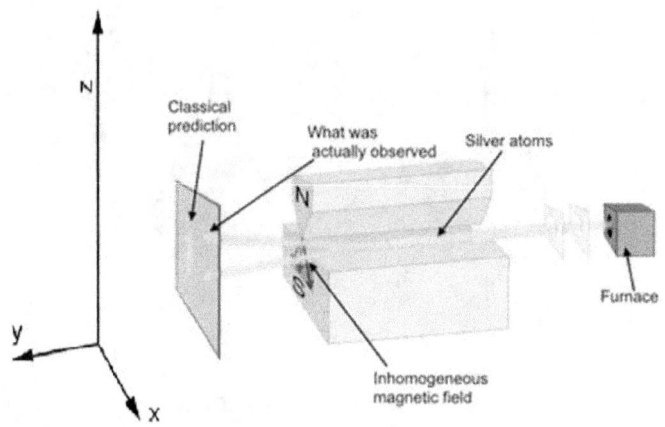

As a result, it is obtained that the atoms, due to the spin of the split electron, deviated along only two opposite directions, as an obvious expression of the quantization state of the measured quantity.

Otherwise, if the Spin number did not have quantized but continuous values, according to a classical prediction, the atoms on leaving the magnetic field should have deflected to all possible positions, including intermediate ones, between the maximum values.

This phenomenon clearly highlights the quantization of Spin's quantum state.

Thus in the z-axis component the above S-vector, can take only upward and downward values, called the up and down directions, respectively, such that the values of the two possible spin moment components are equal to

$$S_u = +\frac{1}{2}\hbar \ , \ \ S_d = -\frac{1}{2}\hbar$$

Graphically, spin states are represented as follows:

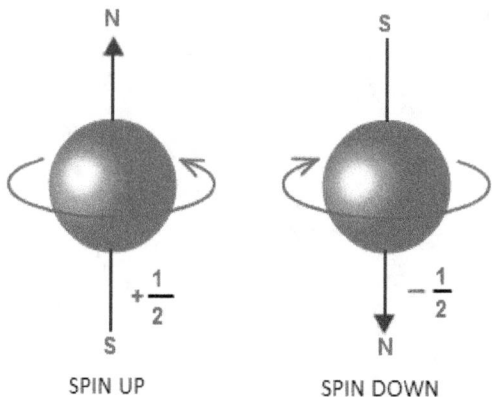

SPIN UP   SPIN DOWN

Spin in a simplistic way, in an attempt to assign a classical interpretation, recalls the rotation of the charged particle around its own axis and can be expressed as that value indicating the number of revolutions the particle will have to perform in order for it to show the same face again.

Continuing by analogy, consider a planet, rotating on itself, such as the Earth for example.

After a full circle completed by the planet how many times will the face of origin be shown?

The answer is "one," one time after a 360-degree rotation.

Let us now take, a coin with two equal Head-to-Head faces.

The coin will show me the Head face after a 180° rotation, and again the Head face after the 360° rotation, in total it will show me the required face twice in the course of a full rotation.

This value, if the planet and the coin were quantum particles, would represent the associated spin.

Based on the previous examples, the planet has Spin 1 and the T-T coin has Spin 2.

For particles, however, we are not talking about faces to be shown again, but about the direction of rotation.

A particle with Spin ½, means that after a 360° turn it cannot show the same initial direction again, but will need as many as 2 turns. It can also be said that such a particle in a full spin only half shows the initial direction.

A spin 0 (zero) particle, on the other hand, means that after one complete spin it shows zero times its direction of rotation, and so it will also be after infinite spins, and thus the spin 0 particle, does not change its direction of rotation.

We can interpret fractional spin also, as if the spinning electron moves lying on top of a Möbius. tape, such that it shows the same direction of rotation after 2 revolutions.

The measurement of spin state value has an important application in the medical field through MRI (Nuclear Magnetic Resonance Imaging).

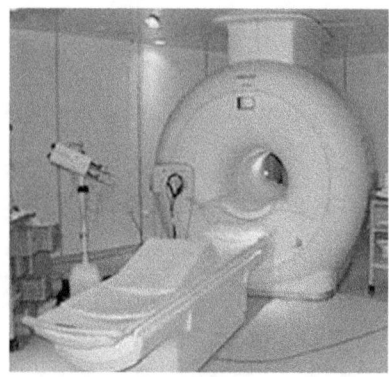 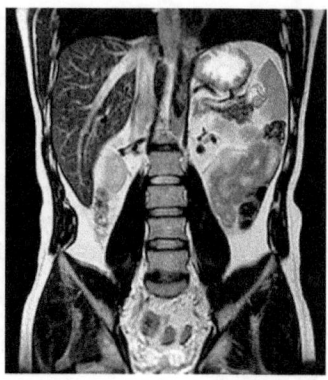

Specifically, the process is carried out by subjecting matter to a special magnetic field.

Under such conditions the protons, or in general the nuclei of the atoms constituting matter, acquire a precession of their spin, like a spinning top.

From the measurement of the spin precession value, the morphology of the biological body can be reconstructed.

MRI is called nuclear, only because it intervenes on the measurement of nuclei properties, and it is absolutely harmless unlike other radiological type techniques, although the adjective nuclear might instill some fear.

The only caution to be used when performing MRI is not to introduce metal objects, including internal objects such as pacemakers, metal prostheses (teeth, eyes, bones, etc.) or otherwise interacting with magnetic fields.

COMMONS.WIKIMEDIA.ORG

"It would be most satisfactory if physics and psyche could be seen as complementary aspects of the same reality"

WOLFGANG ERNST PAULI

https://www.goodreads.com/author/quotes/

2.8 PAULI EXCLUSION PRINCIPLE

Spin plays a very important role particularly because its state is responsible for the stability of matter, through the application of the Pauli exclusion principle, formulated by Austrian physicist Wolfgang Ernst Pauli, for which among other things he won the Nobel Prize in 1945.

Pauli was born in Vienna on April 25, 1900. He was called "a little spirit who appears where theoretical physics studies are cultivated." At a very young age he published a review article on the theory of relativity, and that work is still considered a masterpiece of scientific didactics. He  completed his scientific education first in the stimulating atmosphere of Göttingen and then at the famous Copenhagen Institute, where he found in Bohr a teacher and a friend. His discovery, the exclusion principle, formulated in early 1925, became the most important guide for interpreting atomic and nuclear spectroscopy, connected with the structure of matter. Called to the University of Zurich, he remained there, except for the period of World War II, which he spent in Princeton (USA) at the Institute for Advanced Study, until his death. During his time in Zurich, through his hypothesis on the existence of a neutral particle, later called a neutrino, he provided the key to a complete and consistent interpretation of beta decay in the field of radioactivity. He died in Zurich on December 15, 1958.

Pauli's exclusion principle, finds application only for particles classified as fermions (electron, neutrino, quark, proton,etc.), which have half-integer spin and are among the particles that make up ordinary matter. It does not apply to bosons (photon, gluon, etc.), which have integer spin.

According to this principle, electrons, which as seen above can assume only secondary spin with values of +½ or -½, cannot coexist within an orbital at the same energy level with all quantum numbers equal (n, l, m, m_s), so they can occupy the same orbital, same energy level, same shape and orientation, only if they have opposite spin.

spin up and down electrons

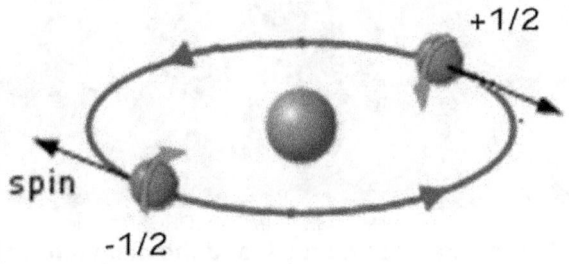

+1/2

spin

-1/2

Given that atoms are composed largely of vacuum, the doubt about the interpenetration of the orbitals of matter is resolved by means of the latter principle.

The component atoms and molecules of matter, cannot be arbitrarily intertwined with each other, because if two electrons with opposite spin are already present in an orbital, no other electrons can be inserted, only secondary spin quantum number values m_s equal to +½ e -½ .

"I Our friend Dirac has a creed; and the main tenet of that creed is: There is no God, and Dirac is his prophet"
WOLFGANG ERNST PAULI
https://www.goodreads.com/author/quotes/

2.9 HEISENBERG UNCERTAINTY PRINCIPLE

With the quantum atom formulation, the orbital is no longer a solid element but consists of and is represented by a probability cloud of electrons.

The electron occupies empty space and lives through a continuous electronic dance in quantum orbitals.

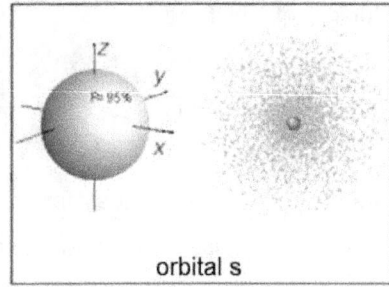

orbital s

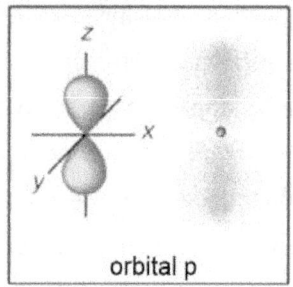

orbital p

With reference to the electron, it is no longer possible to speak of trajectory, as in classical physics, but of probability that it occupies a given space.

By representing the probability density of finding an electron with a given energy level at a given spatial location, we obtain a cloud that the more dense it is the more it indicates another value of probability that the electron is at that spatial location, as best represented by the figures above.

In classical physics, knowledge of the position and forces applied to a body at a particular moment makes it possible to describe the motion of that body at successive moments and thus to know the subsequent position and the ways in which the quantities characterizing the motion vary.

Otherwise, with quantum physics we find the impossibility of being able to know of a particle, simultaneously the value of two conjugate variables, such as for the electron, position and velocity, except within the limits of the uncertainty principle, formulated by the German physicist Werner Karl Heisenberg.

Heisenberg was born in Wurzburg, Germany, on December 5, 1901. During the course of his studies at the University of Munich, of which he later became director, he was a pupil of Arnold Sommerfeld and had Wolfgang Pauli as his desk-mate. After graduation he had the  opportunity to perfect his studies at the two most famous research centers, for quantum mechanics: Gottingen and Copenhagen. When he was only 25 years old he published his famous work on the uncertainty principle. For his studies on quantum mechanics, he was awarded the Nobel Prize in 1932. During the last World War he was one of the heads of nuclear research for the Third Reich, fortunately with mediocre results. He died in Munich on February 1, 1976.

Heisenberg's Uncertainty Principle admits that the simultaneous measurement of two conjugate variables, such as position and momentum, cannot be accomplished without an ineradicable minimum share of uncertainty, unlike classical physics where knowledge of a body's position (coordinates) and momentum defines the future evolution of the physical state under consideration.

However, the condition of indeterminacy does not arise from the lack of knowledge of any hidden variables, but is a characteristic peculiar to matter at the microscopic level, as if it does not want to be observed.

At the atomic level, knowledge of the position and momentum variables of an electron are correlated by an indeterminacy value related to Planck's constant.

Heisenberg formalized his considerations, through a thought experiment inherent in the problem of finding the exact position and momentum of an electron, using a microscope, which instead of using visible light, uses radiation of an appropriate wavelength, compatible with the size of the observation to be made.

In order to interact and thus measure an electron, it is necessary to use radiation with small wavelength values, having comparable equivalent mass.

But a photon having a small wavelength value and therefore a high frequency, due to the Compton effect, interferes with the particle to be observed, changing its momentum.

As a result, the exact position of the electron will be found, but the momentum values will be changed in the collision process.

In practice according to said principle, the indeterminacy of two conjugate variables, such as position and momentum, is not due to lack of information or instrumental inaccuracy, but is a characteristic peculiar to the microcosm that tries to congenitally and naturally oppose observations of its own behavior.

In analytical terms, Heisenberg derived an inequality, which in modern terms is reported as follows:

$$(2.9.1) \quad \Delta x \, \Delta p_x \geq \hbar$$

Dx=statistical uncertainty of the electron position

Dp=statistical uncertainty of momentum measurement (p = velocity x mass)

$\hbar$ = Dirac constant or reduced Planck constant = $h / 2\square$

The indeterminacy relation between position and momentum, was derived by means of a mental experiment, performed by Heisenberg himself and reported in a publication of March 23, 1927, expressed in an initial formulation little different from (2.9.1).

We report a simplified explanation of the experiment.

Consider a photon coming along the horizontal x-axis, when it bumps into a stationary electron, due to the Compton effect, the electron will begin to move and the photon changing direction, will deflect toward a microscope, varying its initial frequency or wavelength λ, so that the final wavelength λ' will be greater than the initial one.

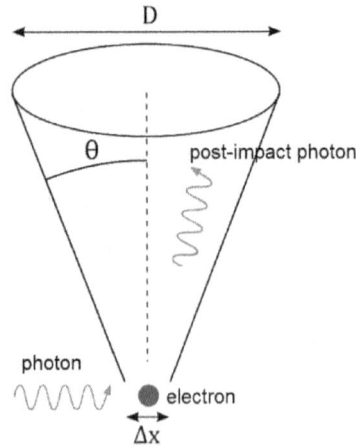

Assuming that the optical microscope has an angular acceptance equal to θ, we can obtain the value of the optical resolution □x with which the microscope observes the electron. The quantity Δx representing the resolution of the microscope also describes the uncertainty of the position of the electron along the x-axis, which according to Abbe's criterion takes a limit value proportional to the following quantity

$$(2.9.2) \ \Delta x \approx \frac{\lambda'}{2\ sin\theta}$$

Taking the axis of the instrument as the reference, the momentum of the deflected photon λ' is equal to *p'*, and its component along the *x-axis* varies between *-p_x '* and *+p_x ' as a* function of the degree of broadening of the optical beam θ, such that a maximum indeterminacy value of

$$\Delta p_x \approx 2\ p'\ sin\theta$$

Recalling that the momentum of a photon is expressible in terms of wavelength, substituting (2.2.4) gives

$$(2.9.3) \ \Delta p_x \approx \frac{h}{\lambda'} 2\ sin\theta$$

Multiplying (2.9.2) with (2.9.3) gives the value of position and momentum indeterminacy

$$\Delta x \cdot \Delta p_x \approx \frac{h}{\lambda'} 2\ sin\theta \cdot \frac{\lambda'}{2\ sin\theta} \approx h$$

The result obtained, gives us a first semi-quantitative estimate about the indeterminacy of two variables proportionally conjugated to a discrete value.

The indeterminacy relation, later in 1929, again by Heisenberg, takes the following form

$$\Delta x\ \Delta p_x \geq \hbar$$

Where $\Delta x\ \Delta p_x$ represent the average indeterminacy of position and momentum, respectively.

In other treatments, through the application of the mathematical formalism of quantum mechanics, it is also possible to find the following relationship

$$\Delta x\ \Delta p_x \geq \frac{\hbar}{2}$$

Where Δ denotes the uncertainty or in other cases the mean square deviation or in others the standard deviation, relating to the measurements of the conjugate variables.

The indeterminacy relation generalizes the concept that all phenomena at the atomic level are equally describable through the theory of classical mechanics, accompanied, however, by intrinsic indeterminacy when investigating two canonically conjugate quantities.

As a direct consequence of the uncertainty principle, it is therefore no longer possible in quantum mechanics to speak of the trajectory of an electron, the position of the electron being expressible only in probabilistic terms.

Heisenberg's theory, developed in the year 1925, constitutes the first formalization of quantum mechanics through a mathematical theory based on the use of matrix mechanics.

Matrices are algebraic elements, which have the peculiarity that they do not respect the commutative property; in particular, a matrix A multiplied by a matrix B is different from the result of a matrix B multiplied by matrix A.

Matric calculus was quite difficult and unfamiliar at that time, while it was easier to work with differential calculus, referring to elements of a continuous type.

So much so that almost simultaneously and independently the formulation of quantum physics through differential calculus was being developed by the Austrian physicist and mathematician Erwin Schrödinger, which we will elaborate more on in the following section.

"What we observe is not nature itself, but nature exposed to our method of questioning"
WERNER KARL HEISENBERG

2.10 WAVE FUNCTION - SCHRÖDINGER EQUATION

Austrian physicist and mathematician Erwin Schrödinger, concurrently with Heisenber and independently, approached the formalism of quantum physics from a wave perspective, introducing a wave function $\psi(x_i, t)$.

Initially, the wave function $\psi(x_i, t)$ represented the time evolution of one or more quantum states of a system (electron, atom, etc.), in the nonrelativistic limit, that is, without taking into account the deformations of the variables as a function of the velocities of the particles in the system, as predicted by the theory of special relativity.

When considering the wave function in the position variables alone, its squared modulus is related to the probability of finding a particle in a given spatial region , in analogy to the wave theory of light, for which the square of the amplitude of the light wave in a region represents its intensity.

With reference to atomic composition, the wave function becomes representative of the indeterminate position of the electron and additional state variables.

Schrödinger was born in Vienna on August 12, 1887. Fostered by his father's cultural sensitivity, he devoted himself from his early childhood to studying the humanities and sciences and learning the major foreign languages. After graduating from the University  of Vienna, he embarked on a brilliant academic career that took him from Vienna to Stuttgart, Zurich and Berlin. After the advent of Hitler, despite his Catholic background, because of his aversion to Nazism he left Berlin to continue his work as a lecturer at Oxford and then Dublin. In 1956 he returned to his hometown to teach until the last days of his life. Bohr called him a "universal man," as a scientist with multiple cultural interests: from philosophy to physics, from history to politics, from biology to Greek culture. A man characterized by a widespread contempt for conventional morality. He combined a deep pessimism with a voluptuous indulgence in the pleasures life could offer. Einstein called him an "overly intelligent libertine scientist," summarizing his virtues and weaknesses. For his equation, he shared the Nobel Prize in 1933 with Dirac, who generalized the corresponding equation taking into account relativistic predictions. Finally, Schrödinger should be remembered for his solution of some biological problems. His lectures, now definable as molecular biology, were collected in a volume entitled "What is life," published in 1944 when he was teaching at the School for Advanced Studies in Dublin. He died in Vienna on January 4, 1961.

The Schrödinger equation, formulated in 1925 and published in 1926, in a more general way, is a differential equation, where the wave function $\psi(x_i, t)$, which represents the state of the physical system under consideration, is its solution.

Said equation, was formulated from the studies on wave-particle dualism performed by French physicist-mathematician-historian Louis-Victor Pierre Raymond de Broglie.

 De Broglie was born in Dieppe on August 15, 1892. Of noble French lineage, he first devoted himself to literary studies, earning a bachelor's degree in history and law in 1910 when he was only 18; later influenced by his older brother Maurice, a talented experimental physicist, he was attracted to the physical sciences. He became especially interested in the theories, connected with quantum physics, by which Einstein had succeeded in interpreting the photoelectric effect. He developed in organic form the original idea of extending wave-corpuscle dualism to particles in his doctoral thesis in 1924. This work can be considered the starting point of wave mechanics. Appointed professor of theoretical physics, he taught from 1928 to 1962 at the University of Paris. In 1929, at the age of 37, the student prince became the first physicist to be awarded the Nobel Prize for his doctoral thesis, for his discovery of the wave nature of the electron. A tireless worker and scholar, in celebrating his eightieth year he had this to say, "*to consider the last ten years spent as the most scientifically valuable of his life.......... to have understood, beginning at the age of seventy, many more things than before, and the joy one feels is greater than that of lost youth.*" De Broglie died in France in Louveciennes on March 19, 1987.

De Broglie's hypothesis, formulated in 1926, states that typical wave properties are associated with each particle.

De Broglie, formulates his hypothesis from the analogy of the behavior of matter with the description of electromagnetic fields as the solution of Maxwell's equations.

For monochromatic light in vacuum, which propagates along a direction identified by the wave vector $\vec{k}$, the electromagnetic fields result described by the following

$$\phi(\vec{r}, t) = A \cdot e^{i(\vec{k}\vec{r} - \omega t)}$$

where $\omega = 2\pi \nu$ is the angular frequency expressible as a function of frequency ν, A the amplitude of the electric or magnetic field, i the imaginary number, t the time, $\vec{r}$ is the distance vector from the origin

Using Planck's energy quantization law and Einstein's demonstration of the photoelectric effect, he associated the behavior of a particle with that of a wave having wavelength as a function of mass, as described by (2.4.3).

The particle could then be described as follows through a wave function

$$\psi(\vec{r}, t) = A \cdot e^{i(\vec{k}\vec{r} - \omega t)}$$

Based on de Broglie's findings, Schrödinger derived his own equation, as in the following.

Let us consider a particle free to move and its Kinetic Energy in the nonrelativistic field, that is, at speeds not comparable to those of light, which is worth

$$(2.10.1) \quad E_c = \frac{1}{2} m \, v^2$$

By introducing the momentum

$$p = m \, v$$

By isolating the value of v and substituting it into *(2.10.1)* in the absence of external force fields, it is possible to write the total energy as equivalent to the kinetic energy alone

$$(2.10.2)\ E_t = \frac{1}{2}\frac{p^2}{m}$$

The momentum of a quantum object can also be expressed as a function of de Broglie wavelength, as derived in the previous section with *(2.4.4)*

$$(2.10.3)\ p = \frac{h}{\lambda}$$

By introducing the wave number, defined as.

$$k = \frac{2\pi}{\lambda}$$

(2.10.3) can be written as follows.

$$(2.10.4)\ p = k\,\hbar$$

Substituting *(2.10.4)* into *(2.10.2)*, we obtain the expression of the total energy of a particle in the absence of an external force field

$$E_t = \frac{1}{2}\frac{k^2\hbar^2}{m}$$

Posing

$$(2.10.5)\ \omega = \frac{1}{2}\frac{k^2\hbar}{m}$$

You get

$$(2.10.6)\ E_t = \hbar\,\omega$$

Having reached this point, introducing the wave function $\psi(x,t)$, in de Broglie's wave form, for simplicity under the one-dimensional assumption (*x-axis* only) and the absence of an external field and thus external potential (i.e., V(x,t)=0), we obtain

$$(2.10.7) \qquad \psi(x,t) = e^{i(kx-\omega t)}$$

With i equal to the imaginary unit (i^2 =-1)

By partially deriving with respect to time *(2.10.7)* and considering that the variable ω in *(2.10.5)* is independent of time, we have

$$(2.10.8) \quad \frac{\partial \psi(x,t)}{\partial t} = -i\omega \, e^{i(kx-\omega t)}$$

Instead, always deriving *(2.10.7)* but with respect to x, twice, we get:

$$(2.10.9) \quad \frac{\partial^2 \psi(x,t)}{\partial x^2} = i^2 k^2 \, e^{i(kx-\omega t)}$$

Isolating from this only the part related to the exponential function

$$e^{i(kx-\omega t)} = \frac{1}{i^2 k^2} \frac{\partial^2 \psi(x,t)}{\partial x^2}$$

and substituting this obtained function into *(2.10.8)*

$$\frac{\partial \psi(x,t)}{\partial t} = -i\omega \, \frac{1}{i^2 k^2} \frac{\partial^2 \psi(x,t)}{\partial x^2}$$

And again, substituting for ω the value given in *(2.10.5)* and placing on the left the partial derivative of space and on the right the partial derivative of time, we obtain the Schrödinger equation, non-relativistic, partial derivative, relating to the motion of a quantum particle, along the *x-axis* only, in the absence of external potential, thus relating to a free particle

$$-\frac{\hbar^2}{2m} \frac{\partial^2 \psi(x,t)}{\partial x^2} = i\hbar \frac{\partial \psi(x,t)}{\partial t}$$

Under the assumption that the motion of the particle can be immersed in any kind of external potential, however, always considering only one-dimensional motion, Schrödinger formulated the following equation

$$- \frac{\hbar^2}{2m} \frac{\partial^2 \psi(x,t)}{\partial x^2} + V(x,t)\psi(x,t) = i\hbar \frac{\partial \psi(x,t)}{\partial t}$$

Where $\psi(x,t)$ represents the general wave function and no longer the de Broglie wave function.

In the more general case of motion in three dimensions, introducing the Laplace operator

$$\nabla^2 = \frac{\partial^2}{\partial x^2} + \frac{\partial^2}{\partial y^2} + \frac{\partial^2}{\partial z^2}$$

and place

$$(x,y,z,t) = (r,t)$$

we derive the equation in three-dimensional time-dependent Schrödinger space, in the nonrelativistic field

$$- \frac{\hbar^2}{2m} \nabla^2 \psi(r,t) + V(r,t)\psi(r,t) = i\hbar \frac{\partial \psi(r,t)}{\partial t}$$

Generalizing, if we assume that r represents all possible state variables of the system, $r = (x_1, x_2, x_3, x_4, x_5 \dots)$, the wave function $\psi(r,t)$, is solution of the partial derivative equation in abstract space and describes the evolution of all possible states of a quantum object.

The partial derivative with respect to time $\frac{\partial}{\partial t}$, which represents the time evolution of the wave function, in quantum mechanics, is also denoted by $\hat{H}$, referred to as Hamilton's operator.

Given that the left-hand member of the previous equation takes on the appearance of an Energy, using Hamilton's operator it is possible to write the Schrödinger equation, in n-dimensional, time-dependent space, in the simpler and more elegant form below:

$$E \psi = \hat{H} \psi$$

From a mathematical point of view, again referring to the electron in an orbital, the wave function ψ(r,t) is a complex function of spatial and time coordinates and takes on the appearance of a "probability amplitude," while the square of its absolute value |ψ(r,t)|^2 , that of a "probability density," provided it is normalized to unity.

Probability density represents the probability of finding a particle in a given spatial region, in a given quantum state.

It is also possible to write the modulus of the wave function in another way

$$(2.10.10) \quad |\psi|^2 = \psi \, \psi^*$$

In this case the function ψ^* represents the complex conjugate of the wave function.

Equality is possible in that multiplying a complex number by its complex conjugate always yields a real number.

Recall that, the complex conjugate of a complex number is obtained by replacing the *i value* of the complex number, with the *-i* value.

Considering a generic complex number

$$z = a + ib$$

given

$$i^2 = -1 \xrightarrow{equivale} i = \sqrt{-1}$$

Its modulus squared is always a real number and is worth

$$(2.10.11) \quad |z|^2 = a^2 + b^2$$

The complex conjugate of z is indicated by an asterisk placed in superscript and is worth

$$z^* = a - ib$$

Multiplying a complex number by its complex and conjugate gives

$$z \cdot z^* = (a + ib)(a - ib) = a^2 - i^2 b^2$$

Given $i^2 = -1$, we obtain

$$(2.10.12) \quad z\,z^* = a^2 + b^2$$

Comparison of (2.10.11) with (2.10.12) yields:

$$z\,z^* = |z|^2$$

Which proves equality 2.10.10.

The implementation of the wave function in the quantum atomic model, highlights the "randomness" (probability) character of quantum physics, enhancing the substantial difference from classical physics based on "causality" (cause-effect).

Solving the stationary Schrödinger equation, that is, assuming that ψ(r,t) itself is independent of time, the latter reduces to ψ(r). By graphically representing, in three dimensions, the squared modulus of such a stationary equation $|\psi|^2$, having as known conditions the position of the electron and its corresponding 95% probability, for example, of being in that particular position, as the energy levels vary (n=1,2,3, ...etc.), we obtain the three-dimensional figures representing precisely the atomic orbitals, referred to in Section 2.6.

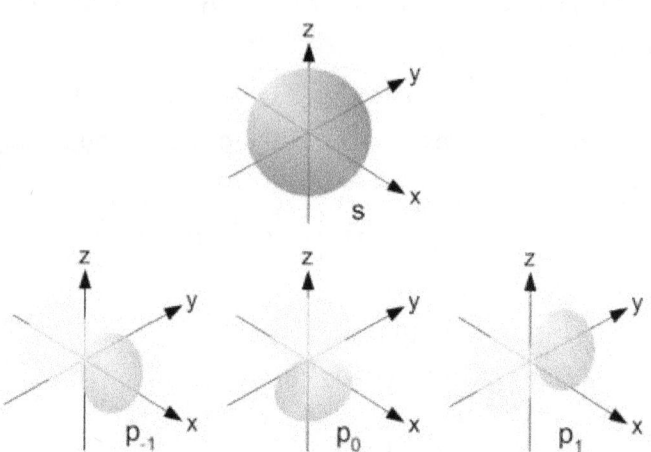

It is evident how simple it is to solve the Schrödinger equation for hydrogen atoms, having only one orbiting electron, while for poly-electron atoms the matter becomes complicated, as electromagnetic repulsion forces between electrons take over. In the latter case, the equation is solved by successive approximations, allowing the atomic orbitals of multi-electron atoms to be represented graphically as well.

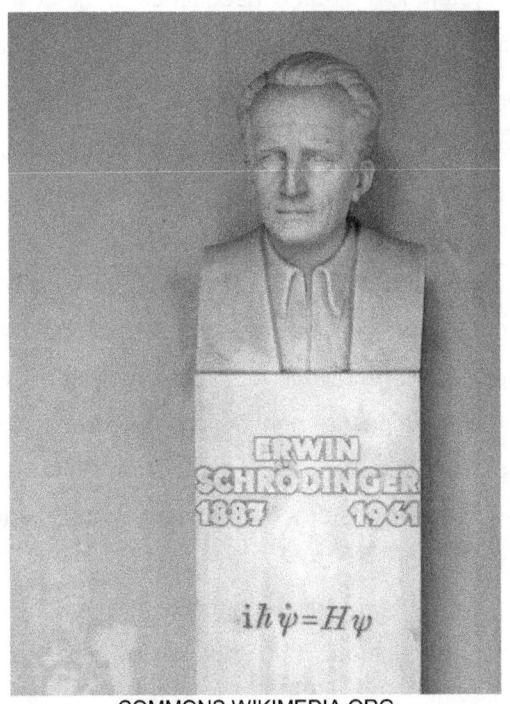

"If a man never contradicts himself, the reason must be that he virtually never says anything at all."
ERWIN SCHRODINGER
https://www.goodreads.com/quotes

2.11 PRINCIPLE OF SUPERPOSITION

A quantum entity exists everywhere in superposition of states until observation is made and the state becomes certain.

From a mathematical point of view, knowledge of the physical state of a system is a direct consequence of the collapse of the wave function $\psi(x,t)$, which causes the reduction of superimposed states into a unique observable state.

An electron can exist both here and there, in any possible state, and it is only when we make the observation that its position becomes certain and the state determined.

Consider, for example, an electron which, as seen above, can have UP or DOWN spin value ($+\frac{1}{2}$ and $-\frac{1}{2}$).

This spin state is not known as long as we make a measurement. As a result of the measurement action, the collapse of the wave function representing the state of the system occurs, so random quantum characteristics are lost to give rise to a certain UP or DOWN state.

Ultimately, the superposition principle tells us that quantum states are not unique, but they enjoy the condition of randomness that also leads to superposed state values in addition to ordinary values.

The superposition principle is an inherent feature of the microscopic world, and in this regard Erwin Schrödinger devised a thought experiment called Schrödinger's cat paradox in 1935, with the aim of illustrating how the interpretation of quantum mechanics (Copenhagen interpretation) gives paradoxical results when applied to a macroscopic physical system.

To this paradox, given its importance, we devote the following section in its entirety.

The concept of quantum superposition becomes clearer by following a surprisingly effective formalism introduced by British physicist, mathematician and engineer Paul Dirac, who in 1927 developed a formalization of quantum mechanics based on noncommutative algebra , similar to Heisenberg's use of matrix calculus, which likewise relies on noncommutative calculus properties.

Dirac was born in Bristol on August 8, 1902. A contemporary of Heisenberg, Pauli and Fermi, he is considered one of the most brilliant theoretical physicists of the century. After graduating from Bristol in 1921, he moved to Cambridge on a scholarship; here, except for a few periods spent in the United States, he always remained, holding, from 1932 until the end of his career, the chair formerly held by Newton. Endowed with a highly analytical mindset and a shy nature, Dirac was known for his extreme reluctance to speak, so much so that his colleagues at Cambridge ironically instituted the "dirac" as a unit for measuring loquacity: "one dirac" was worth the emission of one word per hour. In 1933 he shared the Nobel Prize with Schrödinger for generalizing the corresponding equation, taking relativistic predictions into account. He died in Tallahassee, Florida, on October 20, 1984.

Dirac's formalization, based on noncommutative algebra, involves the use of vectors and operators, as elements of a Hilbert space, capable of representing a quantum state.

Said vectors are not the classical vectors, defined by direction, intensity and direction, and indicated with an arrow in a Cartesian axis system, as we are used to thinking of them in classical physics.

These new vectors, are special in that they are defined in an abstract vector space, and consist of successions of complex numbers or functions of complex numbers, up to an infinite number of components.

We can think of them more as algebraic and not geometric vectors, such that they can be represented through appropriate matrices.

Dirac introduces two fundamental vectors: the *bra-* vector and the *-ket* vector, which together form the word bracket from the meaning: *bracket, group*.

The *bra* vector is denoted by the symbology $\langle A|$ while the ket vector with the mirror symbology $|B\rangle$.

A ket, represents a complex vector that completely describes a quantum state.

The ket vector is representable in an abstract Hilbert space, having special properties of algebraic calculus and in particular characterized by being a complex vector space.

A *ket* enjoys several properties, including some properties of ordinary vectors: they can add together, multiply by a complex (imaginary) number and then combine with each other.

In particular, the combination of two *ket* vectors can be expressed as follows:

$$(2.11.1) \quad a|A\rangle + b|B\rangle = |R\rangle$$

With *a* and *b* two arbitrary complex numbers.

The ket vector $|R\rangle$ being expressed as a linear combination of two vectors ket $|A\rangle$ e $|B\rangle$ is defined as *dependent*.

Conversely, if a ket vector is not expressible as a linear combination of other kets, it is called *independent*.

When a ket vector is dependent it represents an additional state of the system.

In the case of *(2.11.1)*, the ket vector $|R\rangle$ represents an additional state of the system, in addition to the states represented by the vectors ket $|A\rangle$ e $|B\rangle$.

Ultimately a very simple physical system with only two possible states, in quantum terms can have infinite states: the two possible states as a result of the measurement and all possible combinations between the aforementioned states.

Let us examine the famous case of the *spin* state of the electron and apply the above formalism and considerations.

Before performing the measurement, the electron is in a combined state of Spin UP = ψ_1 and Spin DOWN =ψ_2.

Its state can be expressed, adopting Dirac's formalism, as follows:

$$(2.11.2) \quad |\psi\rangle = a|\psi_1\rangle + b|\psi_2\rangle$$

The coefficients a and b, are probability amplitudes that individually only when squared, represent the probability associated with the occurrence of the respective state.

In the analyzed case, with only two possible values (Spin UP and Spin DOWN) we will have

$$(2.11.3) \quad |a|^2 + |b|^2 = 1$$

Which expresses the fact that the sum of the probabilities of obtaining the actual value of a state is definitely par to 100% =1. And again, considering that the electron has equal probability of presenting one of its possible Spins, we obtain:

$$|a|^2 = \frac{1}{2} \; ; \; |a| = \frac{1}{\sqrt{2}} = \frac{\sqrt{2}}{2} = 0{,}707$$

$$|b|^2 = \frac{1}{2} \; ; \; |b| = \frac{1}{\sqrt{2}} = \frac{\sqrt{2}}{2} = 0{,}707$$

And *(2.11.2)* becomes:

$$|\psi\rangle = 0{,}707(\, |\psi_1\rangle + |\psi_2\rangle \,)$$

This relationship tells us that the possible spin values of an electron until the measurement is made are three:

SPIN up, SPIN down and 0.707(SPIN up + SPIN down)

The following part presents mathematical concepts of a higher degree of difficulty, so those who do not wish to try their hand at it can easily move on to read the next paragraph.

Just for completeness, but without going into too much detail, after describing *ket*, let us describe *bra.*

Assigned a set of vectors, it is possible to construct a second set of vectors, called a *dual set* by mathematicians.

A *bra vector is* defined as a vector conjugate to the ket vector, having its elements in a dual space associated with the one given by the *ket*, with the special feature that the scalar products with the corresponding starting *kets* take on assigned values, that is, they are numbers expressed as a linear function of the starting *kets*.

According to Dirac's formalism in representing the scalar or inner product, the symbologies of bra and ket are juxtaposed, eliminating one of the vertical bars and always placing the bra on the left and the ket on the right, so as to form the braket

$$(2.11.4) \quad \langle A| \bullet |B\rangle = \langle A||B\rangle = \langle A|B\rangle$$

The scalar product of a bra $\langle A|$ with the corresponding *ket* $|B\rangle$, as pointed out earlier, is still a scalar, bearing in mind, however, that it does not enjoy the commutative property, such that commuting *(2.11.4)* to a new scalar product having *ket* $\langle B|$ (left) and *bra* $|A\rangle$ (right), instead of a scalar will give rise to a new vector (ket or bra), being

$$\langle A|B\rangle \neq |B\rangle\langle A|$$

In matrix terms, *bra* is denoted as a row vector and *ket* as a column vector .

A tthrough matrix calculation, it becomes apparent how the product between a bra $\langle A|$ (1 row x n columns) and a ket $|B\rangle$ (n rows x 1 column), both associated in a dual space, is a scalar e (matrix *1 x 1* dimensions), since by *(2.10.12)* the product of an imaginary component by its complex conjugate is a scalar.

Be $(a_1^*, a_2^*, a_3^*, ...)$ the components of bra $\langle A|$ and $(a_1, a_2, a_3, ...)$ the components of *ket* $|B\rangle$, we have

$$\langle A|B\rangle = (a_1^* \quad a_2^* \quad a_3^* \quad ...) \begin{pmatrix} a_1 \\ a_2 \\ a_3 \\ ... \end{pmatrix} = a_1^* a_1 + a_2^* a_2 + a_3^* a_3 + \cdots = \lambda$$

In such a matrix representation, the non-switching property can also be checked.

The switched product between the ket $\langle B|$ (1 row x n columns) and the *bra* $|A\rangle$ (n rows x 1 column), is a new matrix at n x n dimensions, i.e., a new vector entity at *n x n* dimensions

$$|B\rangle\langle A| = \begin{pmatrix} a_1 \\ a_2 \\ a_3 \\ ... \end{pmatrix} (a_1^* \quad a_2^* \quad a_3^* \quad ...\,...) = \begin{pmatrix} a_1 a_1^* & a_1 a_2^* & a_1 a_3^* & .... \\ a_2 a_1^* & a_2 a_2^* & a_2 a_3^* & ... \\ a_3 a_1^* & a_3 a_2^* & a_3 a_3^* & ::: \\ ... & ... & ... & \end{pmatrix}$$

Ultimately, the state of a quantum system is expressible through appropriate state vectors subject to a noncommutative algebra in abstract Hilbert space.

These vector entities if they can be measured, either directly through appropriate measuring instruments or indirectly through analytical calculation, are called "observables."

The possibility of indirectly calculating observables through analytical calculus is provided by appropriate linear operators that when applied to the vector representing the physical state quantity, results in a real value, which corresponds to the measure of the physical quantity.

Such linear operators, also defined within the framework of noncommutative algebra in a Hilbert space, are nothing more than special mathematical machines that operating on state vectors produce real values of the quantities sought.

From a mathematical point of view, if O is a linear operator and $|\psi\rangle$ is a state ket vector we have

$$O\,|\psi\rangle = \lambda\,|\psi\rangle$$

Where the value λ is called the eigenvalue and represents the measure of the observed quantity, while the quantity $|\psi\rangle$ is called the eigenvector.

Let us examine the case of the spin operator according to Dirac's formalism.

We have seen how it is possible to perform spin measurements exclusively along one direction.

Consider the z-axis to measure a spin state of an electron that we know can take on the two possible values up and down.

The generic spin state, expressed as ket vectors can be written as a superposition of states, as follows

$$|\psi\rangle = a|\psi_{up}\rangle + b|\psi_{down}\rangle$$

Recalling that the coefficients a and b squared represent the probabilities of obtaining the corresponding state, respectively, conditioned by $|a|^2 + |b|^2 = 1$, then we have that to get the state $|\psi_{up}\rangle$ it is necessary that $|a|^2 = 1$ e $|b|^2 = 0$. By the same reasoning to have the state $|\psi_{down}\rangle$ it is necessary that $|a|^2 = 0$ e $|b|^2 = 1$

The ket vectors corresponding to the two possible up and down states, respecting the algebraic properties of the membership space, can be constructed by placing the coefficients of the superposition relation, on different rows

$$|\psi_{up}\rangle = \begin{pmatrix} 1 \\ 0 \end{pmatrix} \qquad |\psi_{down}\rangle = \begin{pmatrix} 0 \\ 1 \end{pmatrix}$$

Applying to these two states the spin operator defined as.

$$\widehat{S_z} = \frac{\hbar}{2}\sigma_z$$

Where σ_z is the Pauli matrix, which for the z component is worth

$$\sigma_z = \begin{pmatrix} 1 & 0 \\ 0 & -1 \end{pmatrix}$$

And performing the matrix products yields respectively.

$$\widehat{S_z}|\Psi_{up}\rangle = \frac{\hbar}{2}\sigma_z|\Psi_{up}\rangle = \frac{\hbar}{2}\begin{pmatrix} 1 & 0 \\ 0 & -1 \end{pmatrix}\begin{pmatrix} 1 \\ 0 \end{pmatrix} = \frac{\hbar}{2}\begin{pmatrix} 1 \\ 0 \end{pmatrix}$$

$$\widehat{S_z}|\Psi_{down}\rangle = \frac{\hbar}{2}\sigma_z|\Psi_{down}\rangle = \frac{\hbar}{2}\begin{pmatrix} 1 & 0 \\ 0 & -1 \end{pmatrix}\begin{pmatrix} 0 \\ 1 \end{pmatrix} = -\frac{\hbar}{2}\begin{pmatrix} 0 \\ 1 \end{pmatrix}$$

The numerical value present in front of the resulting column matrix is called the eigenvalue and represents the measure of the spin state along the considered axis, i.e., in the case examined it corresponds precisely to the secondary spin quantum number.

The state vector, in such relations, is called the eigenvector, as it is connected to the corresponding eigenvalue.

In this case, the measurements of the spin state along the z-axis component are found to be equal to $\frac{\hbar}{2}$ e $-\frac{\hbar}{2}$ in the spin up and spin down cases, respectively, as expected.

Similarly for the x and y components, the relevant operators are

$$\widehat{S_x} = \frac{\hbar}{2}\sigma_x$$

$$\widehat{S_y} = \frac{\hbar}{2}\sigma_y$$

Whereas the Pauli matrices are worth

$$\sigma_x = \begin{pmatrix} 0 & 1 \\ 1 & 0 \end{pmatrix}$$

$$\sigma_y = \begin{pmatrix} 0 & -i \\ i & 0 \end{pmatrix}$$

Now it is also possible to calculate the modulus of the overall spin state, i.e., the modulus of the spin angular momentum, from (2.7.2), but this time as the sum of the components of the operator

$$S = \sqrt{\widehat{S_x}^2 + \widehat{S_y}^2 + \widehat{S_z}^2} =$$

$$= \sqrt{\left(\frac{\hbar}{2}\right)^2 \begin{vmatrix} 0 & 1 \\ 1 & 0 \end{vmatrix}^2 + \left(\frac{\hbar}{2}\right)^2 \begin{vmatrix} 0 & -i \\ i & 0 \end{vmatrix}^2 + \left(\frac{\hbar}{2}\right)^2 \begin{vmatrix} 1 & 0 \\ 0 & -1 \end{vmatrix}^2} =$$

$$= \sqrt{\left(\frac{\hbar}{2}\right)^2 \cdot 1 + \left(\frac{\hbar}{2}\right)^2 \cdot 1 + \left(\frac{\hbar}{2}\right)^2 \cdot 1} = \frac{\sqrt{3}\hbar}{2}$$

"Pick a flower on Earth and you move the farthest star."
PAUL ADRIEN MAURICE DIRAC
https://www.goodreads.com/author/

2.12 SCHRÖDINGER'S CAT

Certainly it was not Schrödinger's intention to kill a poor cat, for a quantum physics experiment.

The thought experiment is designed solely with the expectations of better understanding the concept of superposition of states, which imposes a strong distinction of the interpretation of quantum phenomena from an interpretation of phenomena according to classical physics.

We place in a closed box a cat, coupled with a diabolical system that randomly triggers a hammer, which can thus break a vial of cyanide.

Upon rupture of the vial, the cat dies. Regarding the baleful end, there is no doubt.

The breakdown of the vial is entrusted to a random event such as the decay of a radioactive substance. For radioactive substances, it is only possible to know the average decay time as a statistic.

We cannot know whether the cat is dead or alive until we open the box and check the cat's health status.

How do we answer the question, Is the cat dead or alive?

The cat is in the superimposed state of alive, dead or alive-dead.

The cat may be in not only the ordinary state of alive or dead, but also in the concurrently alive and dead state, always until the box is opened to perform the observation.

Following the opening of the box, the process of observation involves the breaking down of the "coherence" of the previously isolated system as a result of contact with macroscopic objects placed outside, resulting in the collapse of the wave function and the transformation of the quantum system into a classical system, characterized by certain, or rather "observable," measurements.

For these reasons, in daily life instead of observing quantum-like behavior, matter is posed toward us in a decidedly deterministic manner, precisely because of the effect of so-called "quantum decoherence" or "desynchronization of wave functions."

From a probabilistic point of view, the cat has a 50% chance of being in the Dead state and a 50% chance of being in the Alive state.

In quantum terms, adopting Dirac's notation, indicating the state of the cat with appropriate ket vectors, we can say that the state of the cat is as follows:

$$|STATE\ CAT\rangle = a\,|ALIVE\rangle + b\,|DEAD\rangle$$

For the considerations in *(2.11.2)* and following, the cat having the same probability of being ALIVE and DEAD we have

$$(2.12.1)\ |STATE\ CAT\rangle = \frac{\sqrt{2}}{2}(|ALIVE\rangle + |DEAD\rangle)$$

Which represents the third possible state, following the superposition principle.

Wanting to be more precise, the states should be combined with the condition of the decay of the atom that triggers the cat-killing mechanism, that is, by analyzing the state of the entire ATOM-CAT system.

The atom can be in the decayed or undecayed state with equal probability, so we write:

$$|STATEATOM\rangle = \frac{\sqrt{2}}{2}(|decayed\rangle + |undecayed\rangle)$$

That combined with (2.4.1) and posing:

$$|STATEATOM\rangle = |A\rangle$$
$$|STATECAT\rangle = |C\rangle$$
$$|STATESYSTEM\rangle = |A,C\rangle$$

we get

$$|A,C\rangle = \frac{\sqrt{2}}{2}(|A.DECAYED,G.DEAD\rangle + |A.UNDEC,C.ALIVE\rangle)$$

Ultimately the possible states until the measurement is performed, that is, the box is opened, the cat-atom system is in three superposition of states:

DECAYED ATOM AND DEAD CAT

UNDECAYED ATOM AND LIVING CAT

DECAYED ATOM AND DEAD CAT + UNDECAYED ATOM AND LIVING CAT

"The scientist only imposes two things, namely truth and sincerity, imposes them upon himself and upon other scientists"

ERWIN SCHRODINGER
https://www.goodreads.com/author/

2.13 WAVE-PARTICLE DUALITY

Matter appears to have a dual nature: sometimes it behaves like an electromagnetic wave and sometimes like a solid particle.

This characteristic is expressed in compliance with the principle of complementarity, of which Bohn was a great advocate: "The dual nature of the subatomic world cannot be observed simultaneously during the same experiment."

The dual aspects are complementary, both conceptually and in a physical sense, in that they are mutually exclusive: the observation of wave-like behavior in a single experimental process precludes corpuscular-like behavior.

In understanding this property, an experiment intervenes, called the most beautiful experiment in the world.

The experiment in question is called "of the double slit" and is conducted in analogy to what was performed by British scientist Thomas Young in 1801, the main difference being that the latter

used only electromagnetic waves (light).

Young's experiment is based on the use of a single source illuminating an opaque screen with two parallel slits placed at a small distance and of sufficiently small width compared with the wavelength of the incident light.

YOUNG DIFFRACTION

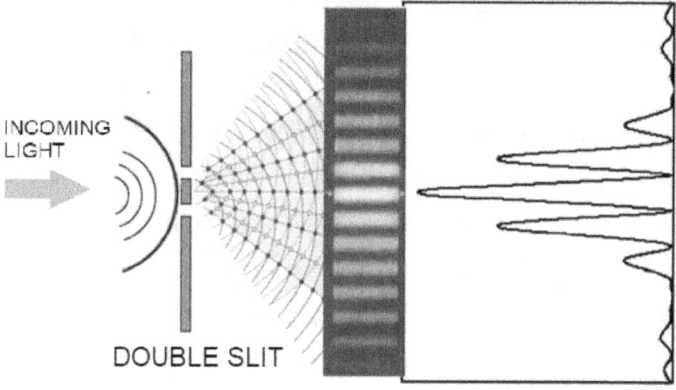

In such a situation, by Huygens' principle, the slits become two linear sources of coherent light that generate on a screen placed at a distance an interference figure formed by alternately dark and bright bands (points of minimum and maximum exposure). Now instead we leave out light and perform the same experiment using matter.

By shooting tennis balls at a screen with two slits, we get that the contact points on the detection plate placed after the slit screen are more concentrated right at the slits.

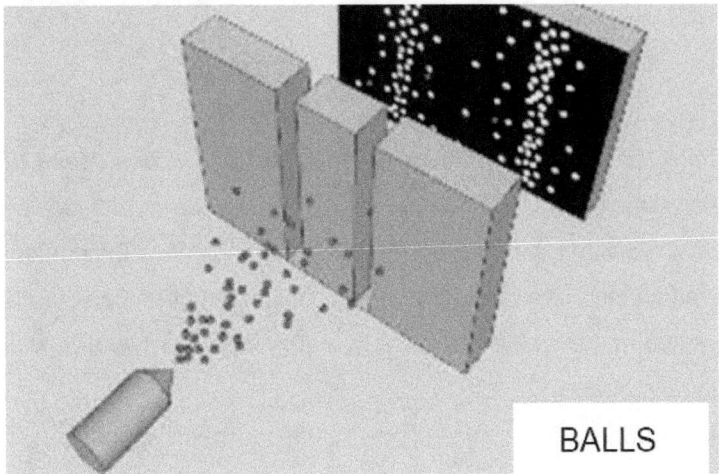

We repeat the experiment by replacing the source-cannon (ball shooter) with a radial perturbation on water, which we know produces waves.

What happens is that an interference pattern with wave peaks and troughs forms on the detection plate, in analogy to the behavior of light.

From the foregoing we could admit that if the experiment is conducted with waves (water or electromagnetic) we get interference figures in accordance with what the wave theory predicts, but if the analogous experiment is conducted using matter (tennis balls) no interference figure is formed and the balls will pass through one of the two slits with equal probability.

If the behavior of the subject matter had been so obvious, I certainly would never have written this book.

In fact, when you switch to performing the experiment with particles at the atomic level, such as electrons for example, something different happens.

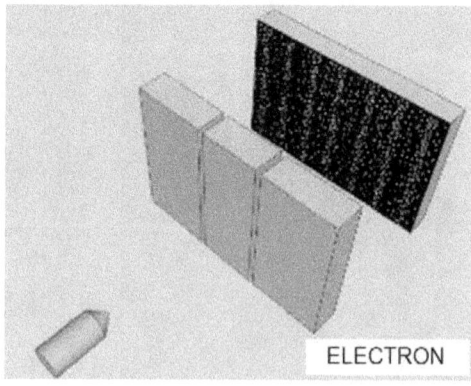

As predicted theoretically as early as Einstein and De Broglie, and verified experimentally in 1927 by physicists Clinton Joseph

Davisson and Lester Halbert Germer, when we treat the experiment with small particles we get an interference figure, hinting at wave-like behavior of matter.

The peculiarity of the above experiment is emphasized by the fact that if the electrons are "fired" individually, the interference figure continues to form.

This is because the motivation for interference does not lie in the interaction of the electron beam with each other, but becomes a characteristic peculiar to the individual electron particle under such boundary conditions.

And again, if one attempts to observe the passage of electrons, with appropriately placed detectors, the interference effect vanishes, as the measurement process causes the wave function representing the particle to collapse.

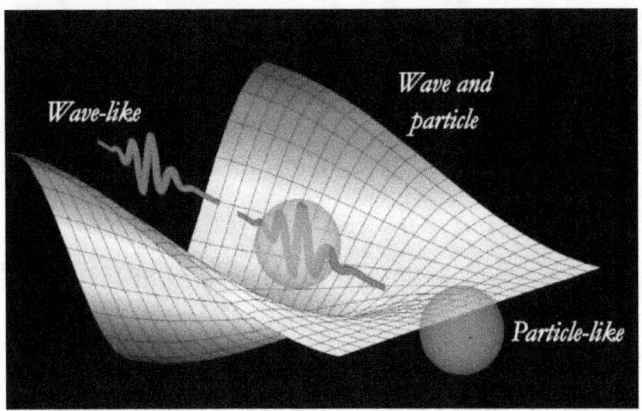

Ultimately, matter sometimes behaves as a wave and sometimes as matter, all summarized as the WAVE-PARTICLE DUALITY PRINCIPLE.

The electron becomes an equivalent wave, represented by its own wave function, which passes through the slits, in the modes proper to waves, and then materializes again on the receiver screen following the collapse of the wave function.

The experiment, can also be interpreted quantumly in terms of the superposition principle, as if the electron passes simultaneously from both the left and right slit, in superposition of state.

Here it is that if you close one of the two slits, what happens is that the interference effect vanishes and the electrons concentrate at the slit, precisely because there is no longer a superposition of possible states.

Still in terms of superposition of states, we can describe the phenomenon with Dirac's notation.

In the case of two slits, the electron can pass into the left or the right one. We have therefore identified two possible states

$$|ELECTRON\ STATE\rangle = a|SX\rangle + b|DX\rangle$$

Due to the considerations in relation (2.11.2) and following, the electron having the same probability of passing through the RIGHT (RIGHT) slot as through the LEFT (LEFT) slot we have

$$|ELECTRON\ STATE\rangle = \frac{\sqrt{2}}{2}(|SX\rangle + |DX\rangle)$$

And thus, the possible states of electrons are LEFT, RIGHT AND LEFT-RIGHT in superposition.

In the case of a single slit, the possible state of the electron is reduced to LEFT and in terms of Dirac formalism

$$|ELECTRON\ STATE\rangle = a|SX\rangle$$

With $|a|^2 = 1$ which equals a probability of 100 percent, corresponding to the certainty that the particle will pass through the only slit present.

"The measure of greatness in a scientific idea is the extent to which it stimulates thought and opens up new lines of research"

PAUL ADRIEN MAURICE DIRAC
https://www.goodreads.com/author/

2.14 QUANTUM ENTANGLEMENT

The word Entanglement, literally translated means "entanglement, correlation."

It is a quantum phenomenon whereby under certain conditions a quantum state of one system turns out to be correlated or intertwined with that of another system, even if placed at a great distance from each other.

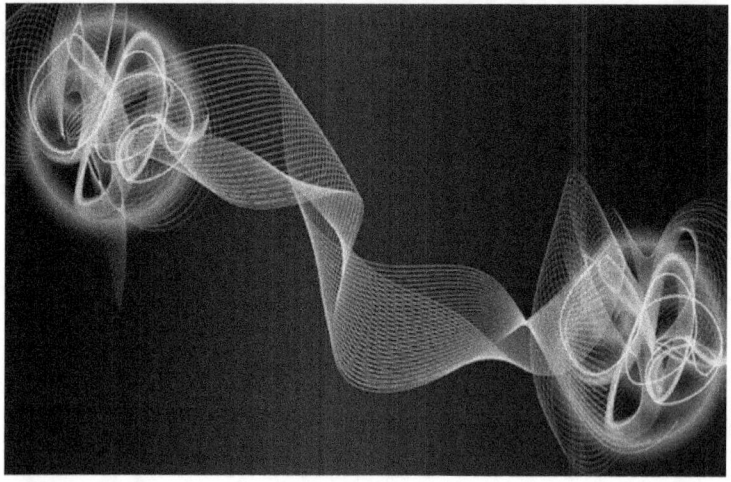

A kind of long-distance correlation, however, which does not travel at the speed of light but is instantaneous.

This phenomenon of so-called "phantom action at instantaneous distance" is in complete contradiction with Einstein's theory of special relativity, which stipulates that the maximum speed attainable is that of light.

Einstein argued that if truly quantum physics was correct, then the world should have been crazy.

Einstein pursued the idea of the existence of hidden, unknown variables, through which the behavior of quantum phenomena could be explained in a "causal" manner, so as to treat even quantum physics with a deterministic character as in classical physics, replacing a probabilistic type of conception in use by the Copenhagen school.

Specifically against the existence of instantaneous remote action, Einstein together with scientists Boris Podolsky and Nathan Rosen, formulated the famous EPR paradox.

This thought experiment was intended to support the validity of the principle of locality, namely, that suitably distant objects cannot have instantaneous influence on each other.

The experiment is to consider two particles that after interacting with each other move away in opposite directions with high but equal and opposite momentum. When the two particles are far enough apart so that they can no longer transmit information to each other at the speed of light, also considering the high momentum, two respective observers make measurements. One observer measures the momentum of the first particle and the other measures the position of the second particle. Given that the particles have the same momentum in opposite directions, it follows that knowledge of the variable position or momentum of one particle implies knowledge of the same variable for the other particle. Consequently, the variables, position and momentum, turn out to be known for both particles, i.e., for the particle system, with extreme precision. The result obtained is paradoxically in complete contrast to what Heisenberg's uncertainty principle states. The consequence is that the principles of quantum mechanics cannot be valid.

Unfortunately, although Einstein never accepted the existence of such instantaneous action he was wrong. Regarding the bizarre behavior according to quantum theory, Einstein was wrong and thus the world was truly shown to be crazy.

The phenomenon of entanglement has been extensively verified experimentally.

The first experimental verification was performed, by probabilistic exclusion, in 1982 by French physicist Alain Aspect, then many more followed.

Aspect by studying the properties of two photons placed in correlation with each other, properly separated and launched in opposite directions, demonstrated the violation of Bell's inequalities, thus verifying with very high probability the phenomenon of quantum entanglement.

With the same experiment he established the exclusion of the existence of any hidden variables of local character, which could cast doubt on the quantum behavior of the two photons.

Remember that the principle of locality states that distant objects cannot have instantaneous influence on each other.

Bell's theorem, in its simplest form, states that "*No physical theory with **local** hidden variables can reproduce the predictions of quantum mechanics.*"

When Bell's inequalities are violated, then any hidden variable theory must also necessarily be nonlocal, such that instantaneous information is exchanged.

Ultimately, it remains confirmed that the quantum world behaves above any conventional perspective, which moreover is also manifested through the existence of entanglement.

Although instantaneous information exchange has been demonstrated for microscopic particles of the quantum order, it should be pointed out that Special Relativity remains abundantly valid for macroscopic bodies.

Let us now see, to better understand what Quantum Entanglement is, with explanatory examples.

Suppose we have two electrons A and B having the SPIN quantum states initially correlated with each other, entangled (precisely "Entangled").

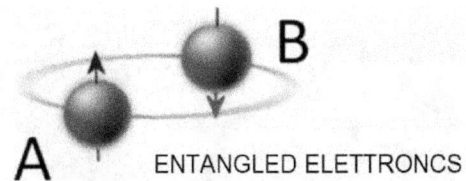

ENTANGLED ELETTRONCS

Electron A will initially have spin UP and electron B will have spin DOWN.

If these are pushed apart and placed at a great distance from each other,

SEPARATION

by performing a change in the quantum state of A (e.g., by varying the spin from UP to DOWN) it happens that instantaneously there is an effect on the quantum state of particle

B. Specifically, the electron B goes from a DOWN to UP spin state, respecting the Pauli exclusion principle.

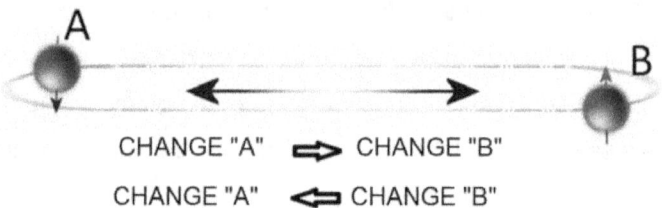

CHANGE "A" ⟹ CHANGE "B"

CHANGE "A" ⟸ CHANGE "B"

This quantum feature is fundamental in studies of quantum computers and teleportation.

Teleportation has always been widely used in the various science fiction films, particularly in the Star Trek science fiction universe.

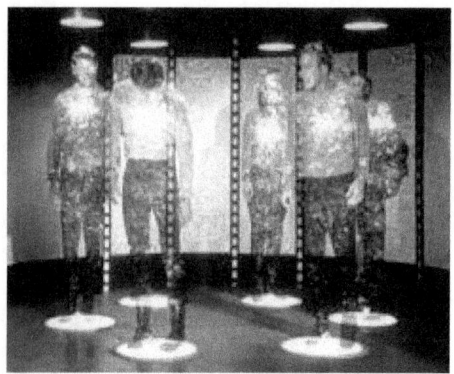

Each of us has always imagined being teleported from one place to another.

In 1993, a group of theoretical physicists tackling the topics of entanglement and non-locality realized that a pair of entangled particles could be used to teleport a quantum state from one location to another distant location, instantaneously, even though the sender did not know the quantum state or location of the receiver, coining the term in "quantum teleportation"

In 1997, just four years after the theoretical discovery, two groups succeeded in the quantum teleportation feat. The first was Danilo Boschi, then at "La Sapienza" University in Rome, and colleagues, followed only a few months later by Bouwmeester's group in Austria, although the latter group published the discovery first.

In 2017, quantum teleportation was demonstrated between a satellite and an earth station in China over distances of up to 1,200 kilometers.

However, one should not confuse teleportation as transport of matter. The phenomenon should be understood as transport of quantum states, which for now is possible with reference only to elementary particles or at most to atoms.

Given, however, that matter is made up of elementary particles, it is not precluded that in the not-too-distant future matter transport could become a reality.

The entanglement property underlies the operation of quantum computers. The super-speed of these computers is mainly related to the characteristic of being able to operate without the constraint of transferring information within circuits at the speed of light.

Quantum computers, in addition to entanglement properties also exploit the principle of superposition, for which the concept of QBIT (QUANTUM BIT) is introduced to replace BITs in classical computers.

In a classical circuit, information is transmitted through BITs that can take only values of 0 or 1, equivalent to OFF and ON.

In quantum terms, the state of a particle, can be not only ON and OFF but also in ON-OFF superposition.

With Dirac's formalism

$$|STATEQBIT\rangle = a\,|ON\rangle + b\,|OFF\rangle$$

$$|0 - 1\rangle = a|0\rangle + b|1\rangle$$

So the possible combinations become endless.

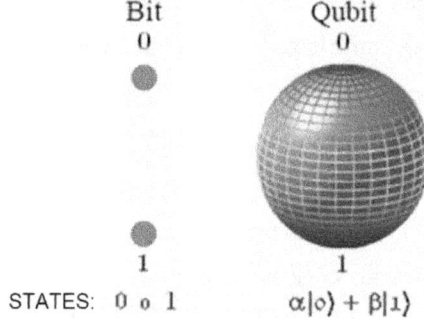

It is since the 1980s that scientists have been trying their hand at developing the quantum computer (or quantum computer), a super-processor that exploits the laws of physics and quantum mechanics to overcome the barriers of today's super-computers and open new horizons for Artificial Intelligence.

Currently, Quantum Computers based on a few QUBITs are already available, and it could still be within a decade at most to the commercialization of full-fledged QCs, given that experimentation and research by IBM, Google, Microsoft, Intel, MIT and Harvard research centers are ongoing.

The following analogy may give a better understanding of the phenomenon of entanglement from a strictly qualitative point of view.

Imagine we have a coin and two cameras pointing at the two sides, respectively.

Room A will observe the face called the Cross, while Room B will observe the other face called the Head.

Now we rotate the coin 180°, so that the Cross face faces B and the Heads face faces A

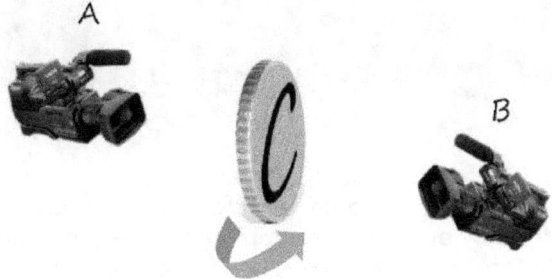

Whenever A changes state from Heads to Tails, as a consequence B also instantly changes state and goes from Tails to Heads.

Thus, we have achieved an instantaneous transfer of entangled or rather Entangled information.

"In science one tries to tell people, in such a way as to be understood by everyone, something that no one ever knew before. But in poetry, it's the exact opposite."

PAUL ADRIEN MAURICE DIRAC
https://www.brainyquote.com/quotes

2.15 OTHER INTERPRETATIONS AND THEORIES

It is useful to mention that there are other numerous interpretations and theories of quantum mechanics, of which only brief mention of the best known will be given in the following.

The many-worlds interpretation, was proposed by Hugh Everett III in 1957 ("Many Worlds Interpretation") and considers the wave function as ontologically real, denying its collapse. Each possibility described by the Schrödinger equation exists in its own specific reality. When we put the cat in the box, the universe splits into two: a universe that contains a dead cat and one that contains a living one. All this implies an almost infinite number of parallel worlds better defined as multiverse.

String and superstring theories, are still developing theories directed toward the unification of quantum mechanics with general relativity (gravity) in order to form a theory of everything. In such a theory, the fundamental constituents are one-dimensional (vibrating) strings, replacing point particles.

The transactional interpretation, abbreviated by the acronym TIQM from the English definition transactional interpretation of quantum mechanics, was presented in 1986 by physicist John Cramer of the University of Washington. It is based on an evolution of the Schrödinger wave equation that takes into account the principles of the theory of relativity (Klein-Gordon equation). This equation contains two solutions describing two waves: a solution describing the flow of energy from the past to the future, delayed waves, and a solution describing the flow of energy from the future to the past, anticipated waves. The transaction between delayed waves, coming from the past, and anticipated waves, coming from the future, gives rise to the well-known wave/particle duality. The wave property is a consequence of the interference of the delayed and anticipated waves; the particle property is due to the location of the transaction.

The statistical interpretation is an extension of Max Born's probabilistic interpretation of the wave function. The wave function is not considered a real entity and is denied application to a single system, such as a photon or particle, while it is imposed that it simply describes the statistical behavior of a set of systems, in the same way that probabilistic laws describe the behavior of the molecules of a gas as a whole. Quantum mysteries are equated with "mysteries" concerning the numbers that might come out of a roll of the dice. Wave/particle dualism does not really exist in this interpretation.

Theories of hidden variables, predicts that quantum mechanics is an incomplete theory, while the behavior of matter remains deterministic in nature and its nature appears indeterminate solely because of the lack of knowledge of hidden variables. Albert Einstein was the greatest proponent of such a theory. But as we have seen, the local hidden variable theory turns out to be incompatible with the results of Bell's numerous inequality experiments, inferring that quantum mechanics would retain its non-locality character.

The de Broglie-Bohm interpretation, ("Guide Wave Interpretation") was originally proposed by Louis de Broglie and later improved and supported by David Bohm. It is part of the so-called "hidden variable" group. According to this theory each type of particle is associated with a wave ("guide wave") that guides the particle's motion.

The pilot wave is very real and permeates the entire universe, constituting its implicate (nonmanifest) order, which Bohm considers to have a hologram structure, in that the global pattern is reproduced in each of its individual parts. What we can observe is only the explicit order, which results from our brain's processing of interference waves. Because Bohm believed that the universe was a dynamic system (whereas the term hologram usually refers to a static image), he used the term "Holomovement" to describe the nature of the cosmos.

In explaining the entanglement process, Bohm states that the two particles or as distinct but interconnected are one at a deeper level of reality. If two different cameras filmed the same fish in an aquarium, in fact, we might have the perception of seeing two strangely  interconnected fish. Any change in the movements of the two fish, in fact, would be synchronous. What in the two televisions (order explicated) seems divided, we know to be a single entity in the aquarium (order implied). Similarly, the two entangled particles would constitute a unity on a more fundamental plane of reality than what our senses perceive.

The interpretation to consistent histories and Ghirardi-Rimini-Weber theory, is a so-called "objective theory of collapse" and introduces the idea that the wave function collapses spontaneously, without any external measurement intervention. Schrödinger's cat is alive and dead for only a very brief fraction of a second and then assumes one of the two states randomly.

Berkeley's interpretation, is based on the notion that the cause of all our perceptions is not an external material reality, but a will or spirit, which was identified with the Christian God; just as the dream is generated by our mind, the universe is a kind of collective dream aroused by God in our souls. Physical reality is not regarded as something existing objectively in and of itself, but only as a mathematical theory existing as a concept in God's mind and projected by God into our minds through the sensory images we perceive; thus both the wave function and its collapse, are real only insofar as they represent the ways in which God conceives of the universe and arouses in us our sensory impressions. This interpretation has no scientific support therefore it is exclusively metaphysical.

COMMONS.WIKIMEDIA.ORG

**"After reading a paper by a young theoretical scientist,
Pauli, shaking his head sadly, commented:
That is not even wrong."**

WOLFGANG ERNST PAULI
https://todayinsci.com/P/Pauli_Wolfgang

3 THE ATOM

3.1 THE SIZE OF THE ATOM

The quantum atomic model, according to the standard model, predicts a central nucleus consisting of neutrons and protons, with neutral and positive electric charge, respectively, surrounded by a probability cloud occupied by electrons, with negative charge, confined in special spatial portions, called orbitals.

Interestingly, in the proportions of the constituent elements of the atom, if the nucleus were the size of an orange then the electron would be the size of a grain of sand, and the radius of the atom would be about 1.00 km.

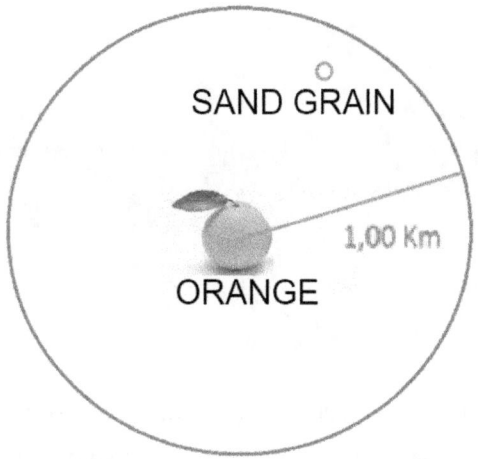

This leads to the consideration that the atom consists mainly, about 99 percent, of the "vacuum," then a small nucleus where almost all the atomic mass turns out to be concentrated, and finally we find tiny electrons.

As a direct consequence it turns out that everything around us, including us, consists mainly of "emptiness" for about 99%.

Since our body also consists mainly of vacuum, how come we cannot pass through opaque components, such as masonry walls?

The motivation lies in the fact that the vacuum, of which the atom is made up, is by no means sterile, but is animated by the quantum dance of electrons in orbital spaces, where atoms themselves cannot arbitrarily interpenetrate, thanks to the Pauli exclusion principle

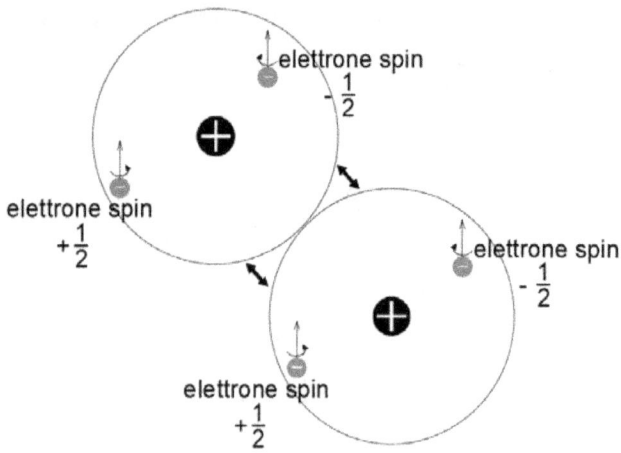

Pauli's exclusion principle states that two electrons cannot coexist at the same energy level and in the same orbital, with the same secondary spin quantum number; given that there are only two possible secondary spin values for the electron: +1/2 and -1/2, the same energy orbital can at most contain two electrons with opposite spins, and therefore the orbitals of neighboring atoms cannot interpenetrate except in the imposed limit.

The particular characteristics of the quantum atom, which have been extensively illustrated in the preceding paragraphs, highlight all the properties that create the distinction from the far-flung concepts of classical physics, and enhance the random and bizarre nature of the atom.

3.2 THE NUCLEUS AND ISOTOPES

The nucleus constitutes the central part of the atom, where most of the mass of the atom itself is concentrated, in view of the small value of the electron mass.

Initially, the nucleus was thought to consist solely of positively charged mass, which balanced the charge of the orbiting electrons.

But under that condition the calculations did not add up, as the theoretical atomic weight turned out to be less than the actual atomic weight.

With the discovery of the Neutron in 1932 by British physicist James Chadwick, the problem of the mass difference highlighted above was solved.

The configuration of the nucleus, consisting of Protons, having positive charge, and Neutrons, having neutral charge, was thus defined.

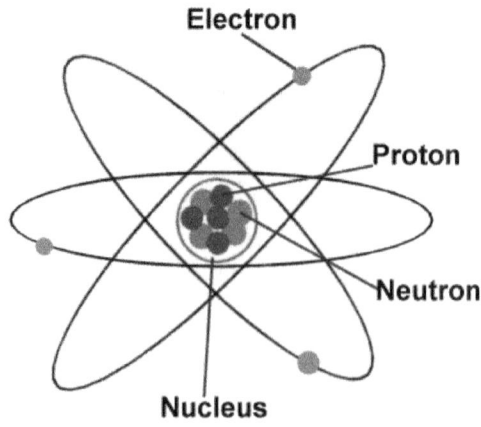

It should be noted that the depiction of the atom as in the above figure is purely figurative, remembering that electrons do not orbit the nucleus, but dance in quantum orbitals.

Ultimately, neutrons are particles involved in defining the mass of an atom, but they are irrelevant in terms of charge and number of electrons.

A chemical element is the heavier the more protons and neutrons its atomic nucleus contains.

Hydrogen, having only one electron and one proton, is a light element. Slightly heavier but still light turns out to be Helium consisting of two protons, two neutrons and two electrons, unlike Carbon which turns out to be much heavier and consisting of six protons, six neutrons and six electrons. And again Iron, evidently the heaviest of the previous elements consisting of twenty-six protons, thirty neutrons and twenty-six electrons. And so on.

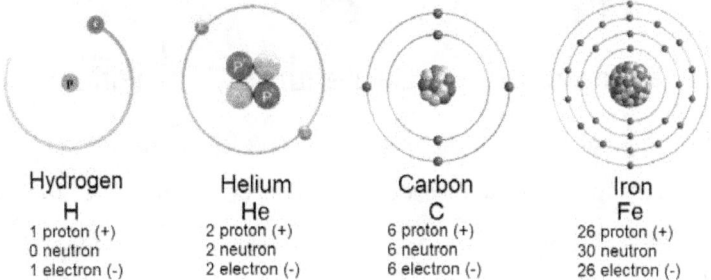

Hydrogen	Helium	Carbon	Iron
H	He	C	Fe
1 proton (+)	2 proton (+)	6 proton (+)	26 proton (+)
0 neutron	2 neutron	6 neutron	30 neutron
1 electron (-)	2 electron (-)	6 electron (-)	26 electron (-)

The same chemical element can have atoms consisting of different numbers of Neutrons, with the same number of electrons and protons. The element thus distinguished is called an ISOTOPE.

Different numbers of protons distinguish the type of elemnto, while different numbers of neutrons distinguish the type of isotope of the same element.

Isotopes are placed at the basis of radioactivity studies.

Let us examine the case of hydrogen as the first element in the periodic table.

Hydrogen can have three types of Isotopes: Common Hydrogen without the presence of neutrons, Heavy Hydrogen or Deuterium with 1 neutron and finally Tritium with 2 neutrons.

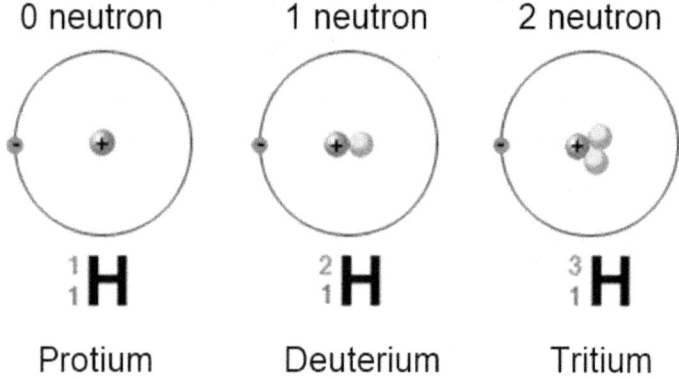

0 neutron   1 neutron   2 neutron

$_1^1H$ $_1^2H$ $_1^3H$

Protium   Deuterium   Tritium

Notice how the number of protons and electrons remain the same, and consequently the total charge also still and always remains neutral.

The neutron number added to the number of protons takes the name "atomic mass number."

$$n_{atomic\ mass} = n_n + n_p$$

Instead, the number of protons, which, due to the neutrality of the atom, will be equal to the number of electrons, represent the "atomic number."

$$n_{atomic} = n_p = n_e$$

Isotopes of the same element will, therefore, have the same atomic number but different atomic mass numbers because of different numbers of neutrons.

Hydrogen isotopes can be summarized by the following table:

ISOTOPO	$n_{atomic\ mass} = A$	$n_{atomic} = Z$
Hydrogen	1	1
Deuterium	2	1
Tritium	3	1

A chemical element is usually identified by two numbers placed before the identifying letter at the bottom and top.

The top number represents in atomic mass number that distinguishes the isotope, the bottom number represents the atomic number that distinguishes the chemical element.

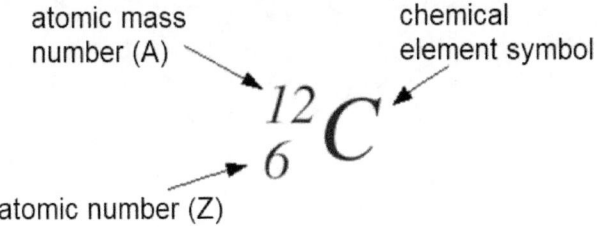

atomic mass number (A)

chemical element symbol

$$^{12}_{6}C$$

atomic number (Z)

4. RADIOACTIVITY

4.1 NATURAL AND ARTIFICIAL RADIOACTIVITY

Radioactivity is a characteristic related to atoms with high atomic number through the decay of unstable nuclei.

When a nucleus has a large number of protons what happens is that the coulombic repulsion force of an electromagnetic nature (of a weak type), overrides the nuclear forces (of a strong type), and in the search for equilibrium the nucleus emits particles that are called "nuclear radiation."

The core then undergoes a "decay" or transformation of the original core through radiation emission.

It is as if within the nucleus the presence of too many protons is unwelcome.

The phenomenon of radioactivity can be natural or man-made.

Natural radioactivity is a characteristic peculiar to some elements of being unstable and decaying over a given, shorter or longer time.

Artificial radioactivity occurs through specially induced processes on certain elements already having certain radioactive characteristics.

Radiation arising from the phenomenon of radioactivity is called alpha rays (α), beta rays (β), gamma rays (γ).

Although they present the designation of rays, they are actually particles.

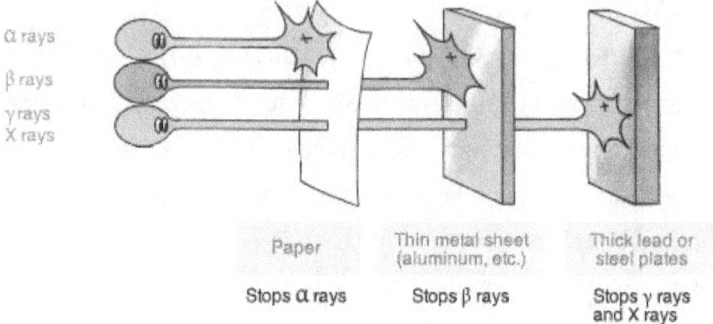

α rays
β rays
γ rays
X rays

Paper

Stops α rays

Thin metal sheet
(aluminum, etc.)

Stops β rays

Thick lead or
steel plates

Stops γ rays
and X rays

α-rays consist of Helium nuclei (2 Protons +2 Neutrons), so they have a neutral-type charge and can be hindered or held back simply by a sheet of paper.

β-rays consist of only one electron (1 and⁻), so they possess a negative charge and can be blocked by aluminum foil.

The rays γ consist of electromagnetic radiation aka Photons. Such photons because of the high atomic energies involved have a high frequency value, by virtue of the proportionality of energy with frequency, given by Planck's well-known equivalence $E = h \nu$.

For this reason, the radiation γ is the most penetrating and dangerous, so much so that lead shielding must be interposed to hold it back.

The three types of radiation are easily separated when placed within an electric field. In such a situation, the rays γ, being uncharged, as they consist of Photons (electromagnetic radiation), will continue undisturbed, **α-rays** will deviate to the negative pole and finally **β-rays** will deviate to the positive pole.

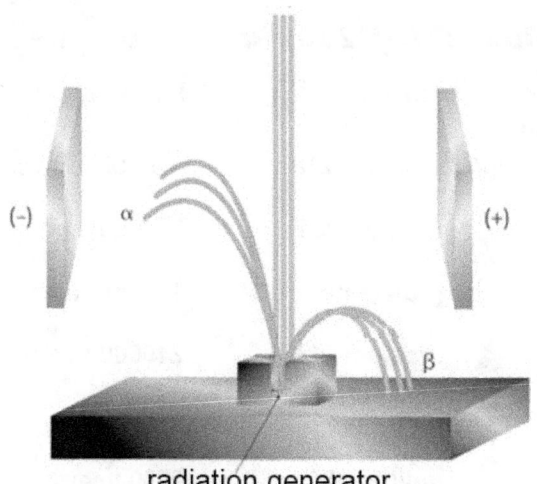

radiation generator

An important measure of radioactive decay is the half-life or half-life, which is a probabilistic measure indicating the time required for half of the atomic nuclei to decay into other atomic nuclei.

Decay times and the type of decay varies with the type of starting element and isotope.

In nature, most elements have stable nuclei or those with fairly long decay times, and this is the case with light atoms (Helium, Hydrogen, Oxygen, etc.), so much so that such elements in nature are stable.

In contrast, elements having heavy nuclei, such as Uranium, Radium, Radon, etc., are unstable in nature and naturally decay into elements having lighter nuclei.

Consider, for example, the decay of the isotope of Uranium type 238.

Initially, Uranium-238 has a mass number of 238, that is, a total number of neutrons and protons of 238.

Uranium 238 (U238) Radioactive Decay

Type of radiation	Nuclide	Half-life
α	uranium-238	4.47 billion years
β	thorium-234	24.1 days
β	protactinium-234	1.17 minutes
α	uranium-234	245000 years
α	thorium-230	8000 years
α	radium-226	1600 years
α	radon-222	3.823 days
α	polonium-218	3.05 minutes
β	lead-214	26.8 minutes
β	bismuth-214	19.7 minutes
α	polonium-214	0.000164 seconds
β	lead-210	22.3 years
β	bismuth-210	5.01 days
α	polonium-210	138.4 days
	lead-206	stable

Following the first stage of decay, it transforms to Thorium-234 through **α-particle** emission in a half-life of 4.47 billion years.

In the second stage, Thorium-234 through **β-particle** emission, in a fairly fast half-life of 24.1 days, decays into Protactinium-234m, which in turn in a half-life of 1.17 minutes, through **β-particle** emission, decays into Uranium-234, and so on we proceed as best illustrated in the figure.

In artificial radioactivity, on the other hand, the decay process is artificially induced, such as by bombarding atomic nuclei with neutrons or nucleons (protons and neutrons) so as to make them unstable.

Included in the latter type are all elements that upon bombardment become elements with atomic number greater than 92, called transuranic because they are artificially obtained. Elements with atomic numbers greater than 109, on the other hand, are called superheavy.

Transuranic and superheavy elements do not exist in nature, except for Neptunium ($_{93}$ Np) and Plutonium ($_{94}$ Pu), which result from the decay of Urbanium-238.

Usually radioactivity is associated with catastrophic events such as atomic bomb explosions, Chernobyl reactor accident, and the use of radiographic instruments in the medical field.

Clearly, the hazard of a radiation is related solely to the interacting quantities.

In fact we are "naturally" and continuously exposed to radiation, indeed radioactive sources are natural components of our bodies.

In a person weighing 70 kg, radioactive elements are present in the following quantities on average:

Carbon 14 (^{14} C) for 12,6 kg

Potassium 40 (^{40} K) per 0,14 kg

Thorium 232 (^{232} Th) per 0,1 mg

Uranium 238 (^{238} U) per 0,1 mg

It clearly emerges that Thorium and Uranium are present in completely negligible quantities. Instead, Carbon and Potassium, which are present in greater quantities than the above, following decay processes release energy that is partly released to the human body through electron generation and partly emitted outward with antineutrinos, through a decay reaction called beta minus (β^-) better specified in the following paragraphs.

Additional radiation to which we are exposed is of the terrestrial and extraterrestrial kind.

Extraterrestrial sources are the stars from which cosmic rays come.

Terrestrial springs are of the natural and artificial types.

Natural sources can be of two types: the first type have been present on the earth since the time of its formation, again through coming from the processes of stellar nucleosynthesis; the second are produced by the continuous processes of interaction between cosmic radiation and atoms in the atmosphere.

Man-made terrestrial sources come from fission processes in nuclear reactors, nuclear explosions, collisions at accelerators in physical and medical research laboratories, and exposure for medical diagnostics (X-rays, CT scans, etc.).

Their average radioactivity is normally lower than natural sources.

Although it is widely believed that radiation has harmful effects, numerous studies have verified that radiation can have beneficial effects in the case of absorption, however, of small doses.

A study reported on the incidence of cancer or congenital malformations in a sample population of 10,000 Taiwanese who were accidentally exposed for 20 years (1983-2003) to 8 to 20 times the natural radiation dose from living in or frequenting buildings constructed using iron accidentally contaminated with the radioactive element Cobalt 60, found that the cases of cancer deaths and congenital malformations were surprisingly and significantly lower than those of the unexposed Taiwanese population, by about 35 times less.

It would appear that small doses of radiation would increase the body's ability to defend itself against cancer and birth defects.

This research alone, however, is not exhaustive and requires confirmation or denial from further studies devoted to the possible beneficial effects of radiation.

4.2 α DECAY

In α decay, one element transforms (transmutes) into another more stable element through the emission of α particles, consisting of Helium nuclei (He = 2 protons and 2 neutrons).

In the case of Uranium-238 having a proton and neutron number of 238, through the emission of a Helium nucleus consisting of 4 protons and neutrons, we have a residue of 234 protons and neutrons, leading to the formation of a new Thorium-234 nucleus.

This type of decay occurs in accordance with the principle of conservation of mass/energy.

INITIAL NUCLEUS	ALPHA PARTICLE	FINAL NUCLEUS

$$^{238}_{92}U \longrightarrow \,^{4}_{2}He + \,^{234}_{90}Th$$

Remember that the number at the top indicates the sum of neutrons and protons, while the number at the bottom indicates the number of protons in the nucleus.

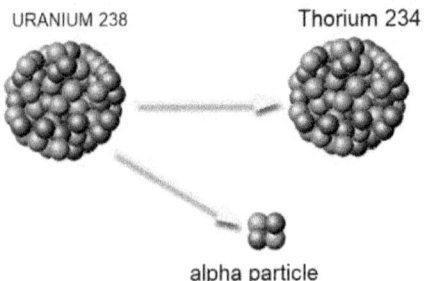

URANIUM 238   Thorium 234

alpha particle

As mentioned above, the obtained Thorium core could be subject to further decay with different half-lives and modes.

4.3 β⁻ DECAY

β decay⁻ (beta minus) is that process by which, in an unstable nucleus, the neutron transforms into an electron (e-), a proton (p+) and anti-neutrino $\overline{v_e}$.

In reality, as we will better elaborate later, it is not the neutron that is transformed but its elemental components, which only as an end result materializes with the transformation of the neutron into a proton, in addition to other particles.

The neutron transformation law is as follows:

$$n \longrightarrow p + e^- + \overline{v_e}$$

It can be seen that a proton is generated from the neutron, which, remaining in the nucleus, produces an increase in atomic number (atomic number = number of protons). In contrast, the generated electron and anti-neutrino are emitted outside.

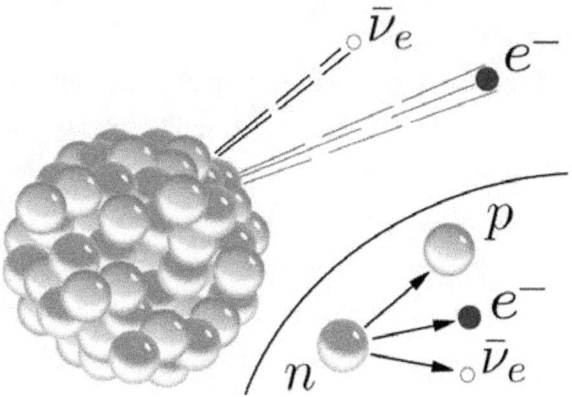

Another new element that appears in the transmutation process is the anti-neutrino, which we will go into more detail later when elementary particles are discussed. For now, let us describe such a particle as consisting of antimatter, with no electric charge, a very small mass of about 25,000 times that of the electron, with spin equal to 1/2 and speed close to that of light.

An example of such decay is given by the Cobalt-60 nucleus, which, following the decay process, transmutes into Nickel-60 by emitting an electron and an anti-neutrino.

$$^{60}_{27}Co \longrightarrow {}^{60}_{28}Ni + e^- + \overline{v_e}$$

The β^- decay reaction becomes of fundamental importance in the process of radiometric dating by the method called Carbon-14 (^{14} C) or radiocarbon dating.

This methodology was devised and developed between 1945 and 1955 by U.S. chemist Willard Frank Libby, who was awarded the Nobel Prize for this discovery in 1960.

In this methodology, we exploit the feature that every living organism has a radioactive carbon 14 isotope component, which decays into Nitrogen 14 (^{14} N), and two stable carbon components ^{12} C and ^{13} C.

Carbon is acquired through continuous exchange with the atmosphere, including through carbon dioxide, through respiration processes or through the nutrition of other living things and organic substances, for the animal world, or through the process of photosynthesis for the plant world.

For this reason, it is possible to radiodate, using the Carbon 14 technique, only things consisting of substances from the plant or animal world (wood, tissue, bones, etc.)

When the organism is alive, the concentration ratio between the isotope ^{14} C and that of the other two stable isotopes ^{12} C and ^{13} C remains constant and equal to the ratio present in the atmosphere.

After death, the organism no longer exchanges carbon with the outside world, and thus the concentration of the unstable isotope[14] C, by decay, decreases relative to the amount of stable isotopes, in a regular manner according to a certain formula.

The decay reaction of the unstable Carbon isotope is as follows.

$$^{14}_{6}C \longrightarrow {}^{14}_{7}N + e^- + \overline{v_e}$$

It occurs over a half-life or half-life of about 5,730 years and according to the following law

$$(4.3.1) \ c = c_0 e^{-\frac{\Delta t}{\tau}}$$

With

c = concentration of[14] C in the organic remains

c_0= concentration of[14] C in the atmosphere

Δt = elapsed time since the death of the organism

τ = average life of[14] C = $\dfrac{halflife \ ^{14}C}{\ln 2}$ = $\dfrac{5.730}{\ln 2}$ = $8.267 \ years$

Through the inverse formula of (4.3.1), knowing the concentration of[14] C present in the organic remains, it is possible to determine the age of the find

$$\Delta t = -\tau \ \log \frac{c}{c_0}$$

However, it is not possible to radiodate fossil finds older than 50,000 years, where Carbon-14 has totally transformed into Nitrogen-14.

4.4 β^+ or INVERSE β DECAY

This process occurs through the transformation within the nucleus of a proton into a neutron (n), a positron (e^+) and a neutrino (ν).

The neutron, generated by the decay process, remains in the new nucleus while the positron and neutrino are emitted outside.

$$p \longrightarrow n + e^+ + v_e$$

mother nucleus

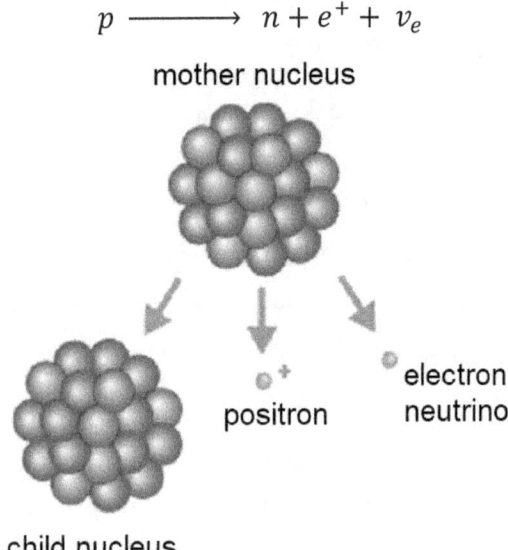

positron

electron
neutrino

child nucleus

In order for this kind of decay to be achieved, it is necessary to provide high energy, at least in the initial start-up phase.

Therefore, such decay results in being typified as nonspontaneous.

The positron is the anti-electron, or the corresponding of the electron as antimatter, and has the same mass as the electron but opposite, positive charge.

The positron has the characteristic that when placed in contact with the electron, materializing the encounter between matter and antimatter, both annihilate in a very short time, about 10^{-9} sec, this annihilates giving rise to two glows, consisting of 2 photons.

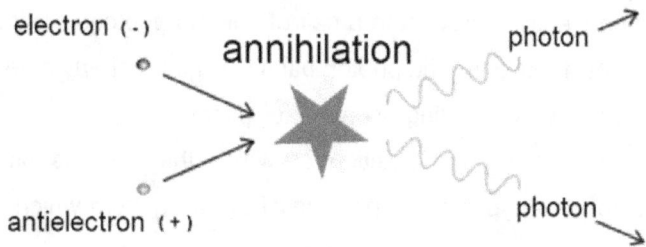

electron (-)   annihilation   photon

antielectron (+)   photon

This particularity of the anti-electron, is used in the medical field in the process called PET (positron emission tomography), which allows, unlike X-rays, physiological type information of matter.

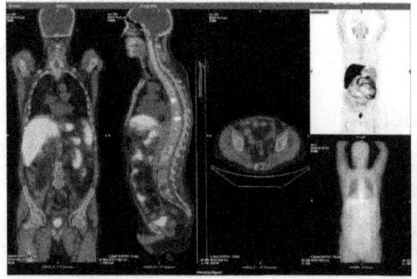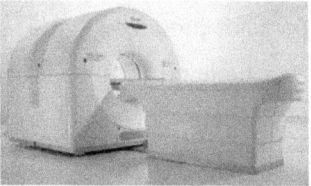

To achieve the desired results, an appropriate process must be followed.

The procedure begins with the injection, to the patient, of a radiopharmaceutical, consisting of a radio-isotope tracer with a short half-life, which chemically binds to a metabolically active molecule (carrier).

The carrier molecule diffuses the radio-isotope into the body to be analyzed.

Because of their low half-life, radioisotopes must be produced by a cyclotron placed near the PET scanner.

The isotope, thus diffused internamnete to the biological body, undergoes reverse β decay by emitting a positron.

After a path of up to a few millimeters, the positron annihilates with an electron, producing a pair of gamma photons emitted in opposite directions. The photon pairs are thus properly detected by a scanner, consisting of photomultiplier tubes.

From measuring the position at which the photons hit the detector, the hypothetical position of the body from which they were emitted can be reconstructed.

This radiological technique produces an irradiation dose equivalent to performing a CT (computed axial tomography) scan, operating with X-rays, and thus equal to about 385 chest X-rays.

Let us return to the decay process and examine the additional particles produced in the reaction.

The neutrino is a particle consisting of matter, which has no electric charge, has a very small mass of about 25,000 times that of the electron, with spin equal to 1/2 and speed close to that of light.

Because of its neutrality and small mass value, the neutrino is a difficult particle to detect.

Fortunately, due to the high velocities possessed by the latter particles, close to the value of the speed of light, it is possible to detect them through the measurement of the corresponding kinetic energy, taking advantage of mass-energy equivalence.

4.5 γ DECAY

This process is not independent but occurs as part of other decay processes, through the emission of photons (rays γ) following the annihilation of an electron (e^-) with a positron (e^+), as described in the previous section.

In terms of reaction

$$e^- + e^+ \longrightarrow 2\gamma$$

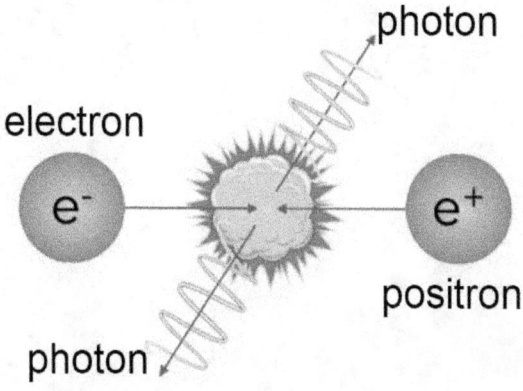

electron

photon

e⁻

e⁺

positron

photon

"Every great and deep difficulty bears in itself it's own solution. It forces us to change our thinking in order to find it."

NIELS BOHR
https://www.goodreads.com/author/

5 NUCLEAR FISSION

5.1 THE CHAIN FISSION REACTION

Nuclear fission is a process of radioactive decay, where the nucleus of a heavy chemical element decays into smaller fragments, emitting a large amount of energy and radioactivity.

A necessary condition for such a process to occur is the presence of a material or rather an isotope of the "fissile" type, this capable of a chain reaction.

An element not capable of following a chain reaction but still divisible is called "fissionable."

For the isotopes of Uranium, for example, we have that Uranium 235 (U^{235}) is a fissile isotope, while Uranium 238 (U^{238}), which is then the most abundant in nature, is fissionable.

To initiate a chain fission process, one proceeds to bombard the fissile isotope with a slow neutron, so that said particle becomes trapped in the stricken nucleus, causing an increase in atomic number so that it becomes even more unstable, until it breaks apart.

The starting core divides into two smaller cores.

The neutron, must have adequate velocity so that it is trapped in the stricken nucleus, otherwise it may pass through it.

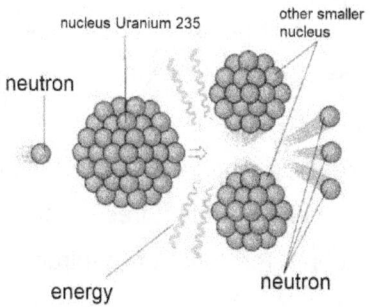

During the fission process, two smaller nuclei, three neutrons and energy are generated for each starting target nucleus and one neutron.

The three neutrons thus generated will be used in the triggering of further fission chain processes.

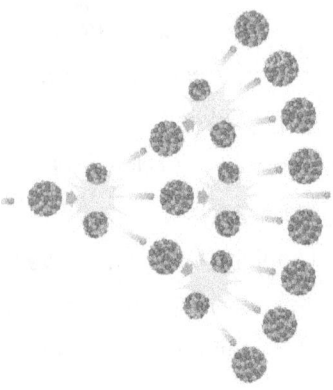

A further condition for the fission process to occur is the presence of a minimum amount of mass, termed "critical mass." The critical mass of a fissile material represents the minimum quantity, which is necessary for a nuclear chain reaction to be self-sustaining.

Let us continue with an example by considering Uranium^{235}. By bombarding the latter element with a slow neutron, such that it becomes entangled in the nucleus, the new configuration will be as follows: a number of neutrons and protons (together called nucleons), altogether 235+1=236, which we could call Uranium^{236}.

The new isotope thus obtained, breaks into Barium^{141} and KRYPTON^{92}, as well as emits three neutrons; altogether the number of nucleons (protons or neutrons) of the starting elements will be equal to the number of nucleons of the generated elements: 1+235 = 141+92+3.

BARIUM^{141}

1 n + U^{235} + 3 n

KRYPTON^{92}

While the sum of the number of starting protons and neutrons is equivalent to the number of outgoing protons and neutrons, what happens is that the input mass is different from the output mass. The difference in the mass of the initial neutron and U^{235} , compared with the sum of the masses of the final products (Ba+Kr+3n) after the reaction, is a consequence of the transformation of part of the mass into energy, due to the equivalence dictated by Einstein's famous relation E=mc^2 .

Given the high value of the constant c, the speed of light in vacuum, it is easy to see how small differences in masses can generate high amounts of energy.

In quantitative terms for a single core of U^{235} you have:

$$E = [(m_{1n} + m_{U^{235}}) - (m_{B^{141}} + m_{K^{92}} + 3m_{1n})]\, c^2 = 211\ MeV$$

with

E = energy developed in the process

m_{1n} = mass of a neutron

$m_{U^{235}}$ = mass of a Uranium-235 core.

$m_{B^{141}}$ = mass of a Barium nucleus 141

$m_{K^{92}}$ = mass of a Krypton nucleus 92

c^2 = speed of light squared

This energy, is manifested by the emission of gamma rays and in a small part (about 5%) is converted into kinetic energy and thus heat.

It is possible to calculate that in a fission reaction with only 16 g of Uranium^{235} there is an energy development of $1.2\text{-}10^9$ KJ equivalent to $3.33\text{-}10^5$ Kwh, which is equal to the energy required to light about 3,300,000 100-watt light bulbs for one hour.

The isotopes Barium^{141} and KRYPTON^{92} resulting from the fission reaction, represent the reaction residues, which in turn, being unstable, further decay producing radioactivity, in the mode of beta decay.

Another advantage of the fission reaction, is that by producing process energy, it is self-sustaining.

A fission reaction, depending on the mode and speed of development, can be uncontrolled or controlled.

Both types of chain fission reaction will be respectively explained in the following paragraphs.

5.2 UNCONTROLLED FISSION REACTION

The fission reaction we have seen to be a self-sustaining chain reaction.

By causing an uncontrolled reaction, an enormous amount of energy is obtained from the process in a short time, through the emission of gamma rays and heat, which is the basis for the making of a nuclear fission Bomb, called the A (Atomic) Bomb, which for our understanding is the "Little Boy" that was dropped on the center of the city of Hiroshima on August 6, 1945.

To make the A-Bomb, it is necessary to have U235, referred to as enriched Uranium, as it is fissile, in an amount of at least 85 percent of the total isotopes.

Uranium in nature is found as isotope 238 for about 99.2%, called depleted Uranium, while as isotope 235 it is found for only 0.72%, other isotopes in small percentages complete the range. The enrichment process assumes the separation of the two isotopes in order to have a higher concentration of U^{235} .

In the atomic arms race, nations enrich naturally occurring Uranium 238 through a long and complex process because of the small mass difference between the two isotopes, which is about 1.26 percent.

For the initiation of the chain reaction, as seen above, it is necessary to reach a mass above the critical mass, called super-critical, but without risking explosion before initiation.

In this regard, the masses are kept separate in blocks of sub-critical masses.

The bomb is detonated with conventional explosives to instantly bring the various sub-critical masses into contact through the collapse of the separators, thus uniting the material in the formation of the super-critical mass.

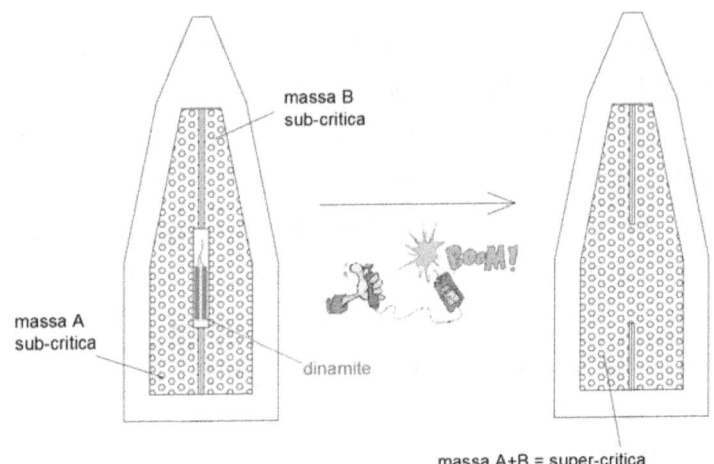

massa A+B = super-critica

There is also a device in the center of the system, containing a strongly neutron-emitting substance, such as polonium, in order to produce the neutrons needed to start the chain fission reaction.

The warhead is possibly coated on the outside with a reflective shield against neutrons that would otherwise be lost to the outside.

The devastating result following the trigger becomes easy to imagine.

$$^{235}U + n \rightarrow ^{236}U \text{ "unstable"} \rightarrow ^{141}Ba + ^{92}Kr + 3n + 211 \text{ MeV}$$

High values of energy in the form of gamma rays (electromagnetic energy), heat (thermal energy) and high particle velocity (kinetic energy) are developed in the chain reaction.

It is precisely the high-energy gamma rays (small wavelength and high frequency), which in addition to the peculiarity of having no mass, permeate all surrounding matter, ionizing it and creating total destruction.

At the same time, the neutron radiation developed in the chain reaction penetrates matter, further altering the composition of the nuclei of biological bodies.

As reaction residues, other unstable isotopes are generated, for that reason subject to further decay, which aggravate the contamination conditions of places even after centuries.

Depleted uranium (U^{238}), on the other hand, being non-fissile, to stay on the subject of military weapons, is used for making ammunition and in the armor of some weapon systems.

When properly treated, depleted uranium becomes as hard and durable as hardened steel and together with the peculiarity of having a high density, when used as a component of anti-tank munitions it is very effective, definitely superior to other much more expensive materials.

" I remember discussions withwhich went through many hours till very late at night and ended almost in despair; and when at the end of the discussion I went alone for a walk in the neighbouring park I repeated to myself again and again the question:?"

WERNER KARL HEISENBERG
https://citations-celebres.fr/auteurs/

5.3 CONTROLLED NUCLEAR FISSION

In the controlled fission process, it is necessary to restrain the speed of the neutrons generated in the process in order to have slow neutrons to achieve a correspondingly slow production of energy, which is transformamile and usable, unlike in the case of the A-Bomb.

The process of controlled chain fission reaction, takes place in appropriate nuclear reactors where placed the fissile material the chain reaction is restrained, or rather moderated, with low atomic number particles.

The first known nuclear reactor is the one built by Enrico Fermi's team in Chicago, in the CP-1 (Chicago Pile 1) reactor, which achieved the first controlled and self-sustaining chain reaction on December 2, 1942.

To moderate the reaction usually heavy water is used, which is water with hydrogen isotopes, such as Deuterium (^{2}H).

In order to be able to slow the neutrons down to a complete halt in the event of the emergence of issues of concern, special control rods, made of suitable neutron-absorbing material, are used.

The rods can be made of silver, cadmium, graphite, or materials with the same neutron-neutralizing characteristics.

A nuclear reactor, in a schematic and simplified manner, can be composed of a central core containing the fuel (fissile material), a moderation zone for slowing down the reaction, a control rod for the neutrons generated, and a zone for the diathermic fluid, which   as   a   result   of   heating

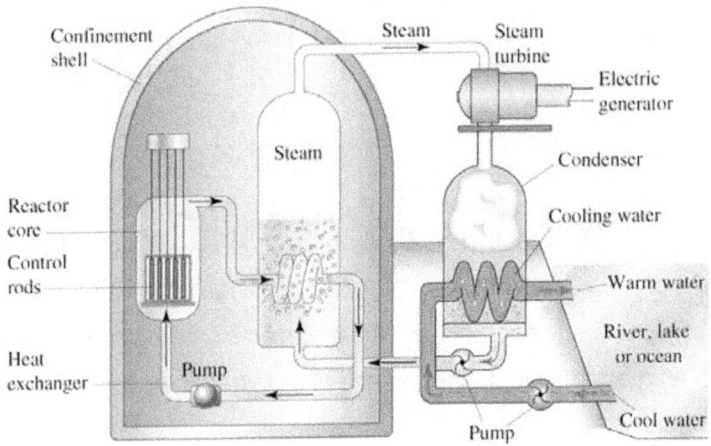

the fluid, drives appropriate turbines for generating electricity.

The dangers associated with the use of fission power plants are well known and due to the risk in the control of the chain reaction, as already happened, among the most recent and serious, in 1986 in Chernobyl (Soviet Union), in 2011 in Fukushima (Japan).

An additional issue, which is not insignificant, in the operation of fission power plants is due to the disposal of waste, composed of radioactive isotopes resulting from the fission reaction process.

"Even for the physicist the description in plain language will be the criterion of the degree of understanding that has been reached."

RNER KARL HEISENBERG
https://citations-celebres.fr/auteurs/

6 NUCLEAR FUSION

6.1 NUCLEAR FUSION REACTIONS

Nuclear fusion is the reaction process between two low atomic weight nuclei, which fuse together to result in a new nucleus with a higher atomic number.

This process is very energy-consuming in the start-up phase, where it is necessary to overcome the electrostatic repulsion forces that are generated between the protons of the corresponding nuclei in the course of their fusion.

Once the reaction, being of the exothermic type, is initiated, there is emission of energy such that the fusion process is energetically self-sustaining; however, this applies to fusion processes of the elements with atomic numbers up to 26 (Iron) - 28 (Nickel) at most.

For all elements with atomic number greater than 28, where the fusion process becomes endothermic (energy absorption), the reaction process becomes no longer energetically self-sustaining.

Let us analyze the case of the nuclear fusion reaction of only two low atomic number elements, such as hydrogen in the isotopes Deuterium and Tritium.

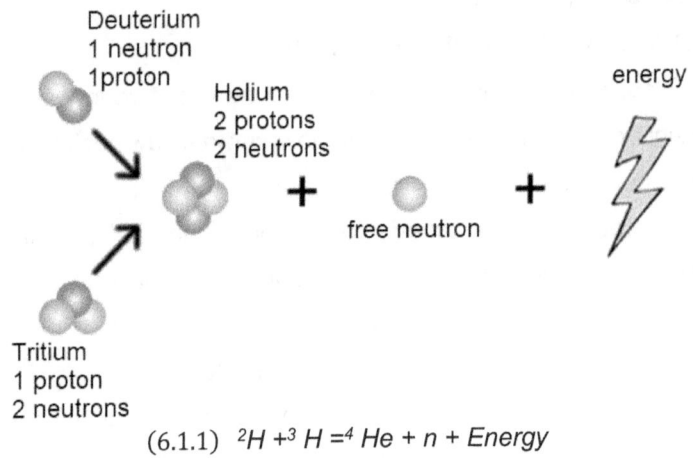

$$(6.1.1) \quad {}^2H + {}^3H = {}^4He + n + Energy$$

From the fusion of a Deuterium nucleus and a Tritium nucleus, a Helium nucleus is generated, in addition to a neutron and Energy emission.

Neutron emission is a problem, which because of its electroneutrality becomes difficult to control with magnetic fields. The Emission of energy of manifested by the so-called "mass defect."

The mass of the starting nuclei is higher than the mass of the nuclei generated as a result of the fusion process.

This difference in mass is a consequence of the transformation of part of the mass into energy, according to Einstein's equivalence $E = mc^2$.

In the case of the fusion reaction of Deuterium with Tritium referred to in *6.1.1,* knowing the values of the respective masses of the reaction elements, we can calculate the energy developed by mass defect

$$E = [(m_D + m_T) - (m_{He} + m_{1n})] \, c^2 = 3,5 \, MeV$$

From this result, taking into account the low value of the atomic weight of the reaction elements compared with the atomic weight of the elements participating in the fission reaction process, the advantage of the fusion reaction over the previous fission reaction is clearly shown.

A fusion reaction actually occurs in several successive stages and in several parallel branches.

Starting from two elements, a new core is formed by combination of the intermediate elements as well.

Let us analyze just one of the possible branches, of a hydrogen nuclei fusion reaction with multi-stage development:

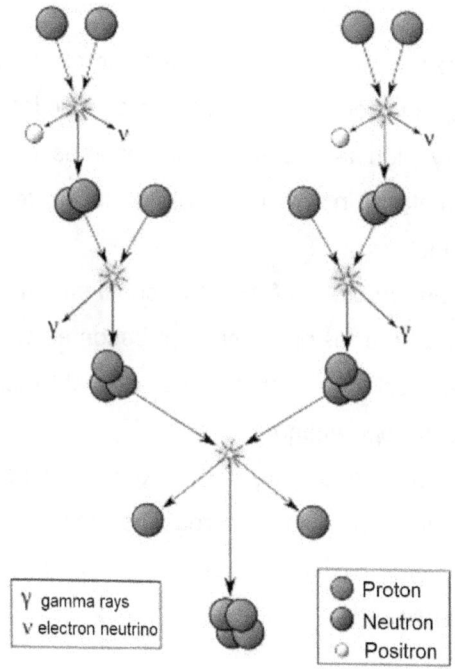

$$(6.1.2) \quad {}^1H + {}^1H = {}^2H + e^+ + \nu_e$$

$$^2H + ^1H = ^3He + \gamma \ + Energy$$

$$^3H + ^3H = ^4He + H + ^{11}H + Energy$$

From the above reactions, it appears that hydrogen nuclei, consisting of one proton, in the fusion process, through intermediate steps and combinations, come together to form Helium nucleus, consisting of two protons and two neutrons, in addition to the production of positrons, neutrinos, gamma rays and Energy.

Nuclear fusion has the major advantage of producing no nuclear waste in the process and producing about ten times more energy than a fission process for the same amount of starting mass used.

On the other hand, given that the process, due to the characteristic of generating exothermic reactions, leads to reaching very high temperatures, it becomes complicated to contain the material in the course of its melting reaction for an adequate time.

In light of the problem of plasma confinement at the high temperatures of the fusion process, it becomes complicated to build a nuclear fusion reactor, so much so that to date there are no such reactors operational.

The only existing facilities are of the experimental type capable of sustaining the nuclear fusion reaction for a very short time through confinement of the fusion plasma by magnetic fields of very high intensity

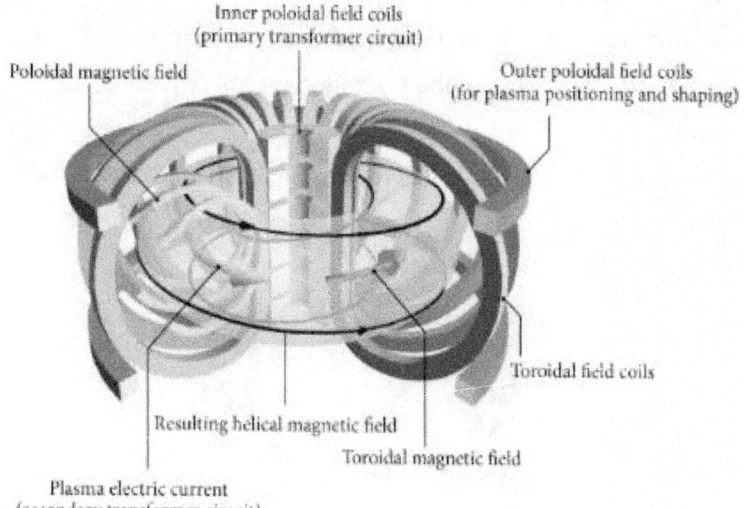

Inner poloidal field coils
(primary transformer circuit)

Poloidal magnetic field

Outer poloidal field coils
(for plasma positioning and shaping)

Toroidal field coils

Resulting helical magnetic field

Toroidal magnetic field

Plasma electric current
(secondary transformer circuit)

It is estimated that the first plants may not be operational until around 2050.

6.2 HYDROGEN BOMB (H-BOMB)

An explosive device that uses the fusion process instead of the fission process is called the Hydrogen Bomb, or rather the H-Bomb.

The fusion process is allowed to occur in an uncontrolled manner with production of high energy values, about 2,500 times that from a similar fission process.

The fuel of an H-bomb is Lithium and Deuterium, and a small fission bomb is used to start the fusion process.

We have already seen that a fission bomb requires a classical device for its detonation.

Then the explosion, through the fusion process, occurs in sequence: detonation Tritol (TNT), fission reaction, fusion reaction.

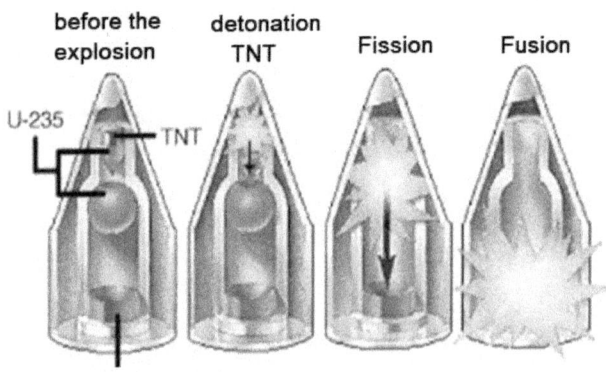

before the explosion   detonation TNT   Fission   Fusion

U-235

TNT

Lithium and Deuterium

"The opposite of a correct statement is a false statement. But the opposite of a profound truth may well be another profound truth"

NIELS BOHR
https://www.goodreads.com/author/

6.3 NUCLEAR FUSION IN STARS

We have seen how the process of nuclear fusion is difficult to replicate in nuclear reactors because of the problem of managing the high temperature values generated during the process.

In stars, on the other hand, as has been the case for millennia, fusion processes, once initiated from the time after the Big Bang, continue spontaneously because of the peculiar exothermic characteristic of the reactions.

The nuclei of low atomic weight elements fuse to give rise to nuclei with higher atomic weights in an energetically self-sustaining process.

Hydrogen nuclei fuse into Helium nuclei, among many others, in the manner seen in the previous paragraph.

In the sun, fusion affects an amount of Hydrogen equal to about 600 million tons per second.

The reaction continues through the fusion of Helium nuclei to give rise to a nucleus with a higher and thus heavier atomic number, and so on.

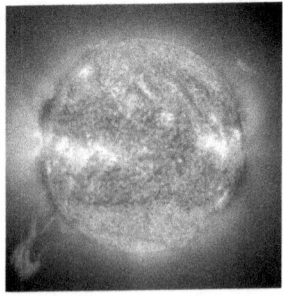

During the process of nuclear fusion of stars, as can be seen from 6.1.2 and following, neutrinos and positrons are emitted in addition to energy and gamma radiation.

Gamma radiation gives brightness to the star, positrons annihilate with electrons in the surrounding space to give rise to more photons, and finally neutrinos continue on their undisturbed path, as they are neutral and of very small mass.

Consider that the Sun emits neutrinos, which reach planet Earth in 8 minutes, with such magnitude that each person results in being invested with a number equal to 10 billion neutrinos per second.

The fusion process in stars continues until the formation of Iron-Nickel nuclei, where the reaction begins to become endothermic and no longer produces process energy.

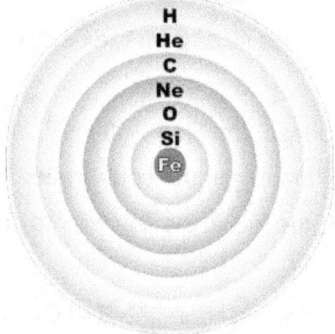

If the star is not massive enough, it cannot generate high values of pressure, to raise the necessary temperature levels, and so it begins to "shut down," as it can no longer energetically sustain the fusion process.

In the case of low-mass stars, such as the Sun, which has an age of about 4.6 billion years, the fusion process continues until the fuel represented by the reserves suitable for "fusing" hydrogen and helium are exhausted.

The term "little massive" is always to be compared with the other stars in the universe, remembering that the mass of the sun is vastly greater than that of the individual planets-in fact, the sun makes up about 99.8 percent of all the mass in our solar system. The mass of the sun is about 1.989×10^{30} kg while the mass of the earth is about 5.972×10^{24} kg, from which running the ratio yields that the mass of the earth is about 0.03% of the mass of the sun.

The sun does not have enough mass to withstand the fusion of elements heavier than helium (He), so the moment the latter "fuel" is about to run out, nuclear/electromagnetic repulsion forces will override gravity and the sun will begin to slowly expand to 20-100 times its current radius, thus becoming a Red Giant.

In the expansion phase, the Red Giant will, among other things, encompass all the planets in the solar system, including Earth.

No worries though: the fuel in the sun will run out in about 5 billion years.

After ejecting the outermost part, the Red Giant will undergo core collapse until it becomes a White Dwarf, and then die out to become a Brown Dwarf.

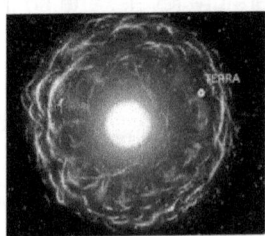 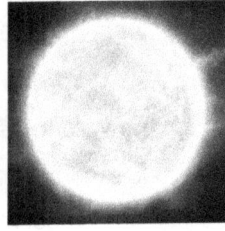 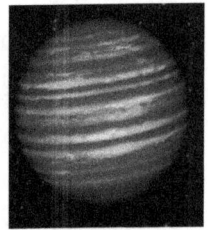

In the case, however, of more massive stars, at least 9 solar masses, gravity prevails over repulsive forces and causes the star to contract, with increased pressures and temperature, so that it is able to melt the innermost layers, at least down to iron.

At the end of the available fuel, unlike in the sun, in the massive star it may happen that gravity tends to concentrate the stellar mass, decreasing its diameter, until it implodes.

The implosion process occurs within seconds, producing the release of shock waves that travel at a speed of 30,000 km/s, or 10 percent of the speed of light in vacuum, and cause an explosion of the star's surface layers.

This explosion process lasts for several weeks and generates extremely high energy and radiation emission, of such magnitude that for short periods it can exceed the brightness of an entire galaxy.

It is bizarre how just in the very last period of its life, before it fades away, this star becomes brighter and more radiant than ever.

At this stage of its existence, the star, takes the name "supernova."

The explosion also involves the diffusion into surrounding space of all the material of which the star was composed, so much so that we can all claim to be "star children," at least by atomic composition.

Following the explosion, a very dense stellar core remains, which depending on the residual mass becomes a Pulsar star if it consists of only neutrons, Quasar if it consists of only quarks (elementary particles constituting neutrons) or in the limiting case becomes a singularity called a Black Hole.

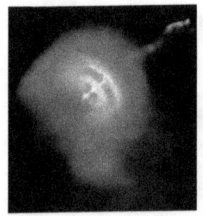  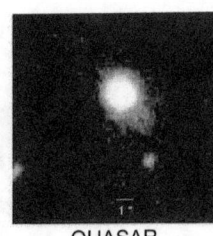 

PULSAR   QUASAR   BLACK HOLE
(GRAPHIC RECONSTRUCTION)

These particular astronomical entities, share a common characteristic of having a high mass density, that is, an enormous amount of mass concentrated in a small volume.

A planet turns out to consist of the bulkiest atoms (central nucleus and orbiting electrons) compared to a dense stellar nucleus consisting of only shaped elementary particles.

And so it is that Pulsars, for example, consisting only of neutrons in the form of plasma, have enormously less volume than a planet with the same mass.

This is obvious if one remembers the atom size considerations performed earlier, where equating the size of the nucleus to an orange, the electron would be the size of a grain of sand and the radius of the atom equal to 1.00 km.

Let us try to derive a qualitative numerical figure regarding the greater mass density of a stellar core compared to that of a planet.

A first result is given to us by relating the average atomic radius value to the neutron radius:

$$x = \frac{10^{-10}}{10^{-14}} = 10.000,00$$

This result indicates that approximately a neutron star, with the same mass, has a radius 10,000 times smaller than that of an equivalent planet.

In terms of mass, considering that volume is a function of radius cubed, we can obtain that for the same size, a Pulsar, compared to a "cold" planet, has a mass ratio of

$$y = \frac{(10^{-10})^3}{(10^{-14})^3} = 10^{12}$$

Ultimately, a Pulsar has a mass about a trillion times greater and that of a "cold" planet, having the same volume and thus the same size.

This huge value of mass is sometimes such that it generates a gravitational contraction of the star, further increasing its mass density.

The high mass concentrated in a small volume can be so high as to configure a singularity, better known as a Black Hole.

The adjective "Black" comes from the fact that even light cannot escape the gravitational pull, such that it is not visible.

The presence of black holes has been ascertained through gravitational studies of the universe, and in April 2019, scientists from the European Commission-funded Event Horizon

Telescope (EHT), with Italy's participation with the National Institute of Astrophysics (Inaf) and the National Institute of Nuclear Physics (Infn), announced the first image of the innermost belt surrounding a black hole.

The "photographed" neo-hole is located at the center of galaxy M87, in the constellation Virgo, 55 million light-years from us, and has a mass estimated at 6.5 billion times that of the sun.

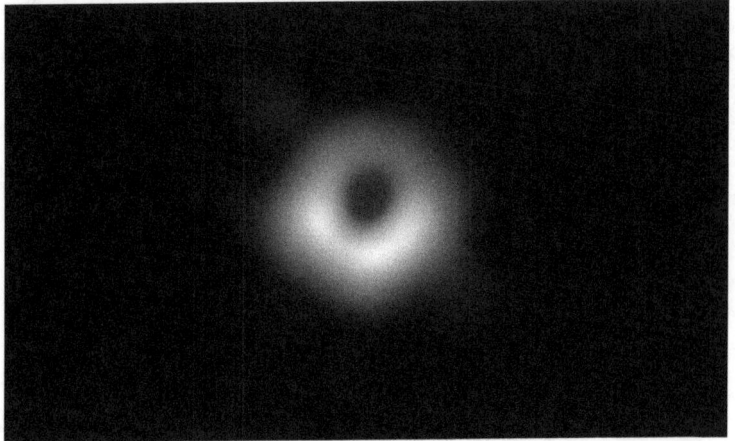

According to the theory of general relativity (see same author's book "The Wonderful Theory of Special and General Relativity - Year 2018") such massive objects also cause time dilation.

If we could, therefore, get close to a Black Hole, time would flow more slowly and upon our return to Earth we would find ourselves catapulted into the future.

Unfortunately, any object that were to approach such a maximal body would undergo a process of spaghettification, that is, it would be attracted radially so strongly as to undo its own structure until it became one-dimensional, just like a spaghetti.

"Those who are not shocked when they first come across
quantum theory cannot possibly have understood it"

NIELS BOHR

7 MATTER AND ANTI MATTER

7.1 THE ANTIMATTER

In the late 1920s, the not-yet-30-year-old Paul Dirac, engaged in the study of quantum theory at high energies, and thus in the relativistic regime, discovered the existence of a particular new particle of opposite charge to that of the electron, which would later turn out to be precisely a particle of antimatter.

Initially Dirac assumed that such a particle was a proton.

Only later, in 1932 Carl Anderson, a young physicist at the California Institute of Technology, was able to provide concrete evidence for the existence of antimatter, and the following year Patrick Blackett and Giuseppe Occhialini completed the discovery, confirming the theoretical prediction of the existence of an antiparticle of the electron.

Anderson's discovery occurred in the course of an experiment designed to study the nature of Cosmic Rays, or the stream of particles from space that strikes our planet at every instant.

The results were obtained by studying the traces left by these particles as they passed through a fog chamber.

The fog chamber, now replaced by more technologically advanced systems, was an apparatus consisting basically of a chamber filled with vapor, in which the passage of a charged particle, by ionization, is visualized by

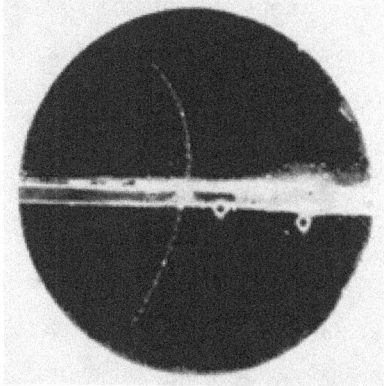

the appearance of a trail of bubbles, like the wake left by airplanes.

Among many ordinary tracks, Andersen identified a particular one, which corresponded to the passage of a particle that deflected in the opposite direction to the electron. Based on the deviation it underwent, the new particle must have had an electric charge opposite to the electron, which, however, could not be a proton because of its size.

Such a particle turned out to be precisely the anti-electron, which he named positron, because of the characteristic of having the same mass as the electron but opposite charge.

With this latest discovery, everything that occupies the universe will turn out to consist of matter, anti-matter and vacuum, not forgetting that by enforcing the equality matter and energy we could add, why not, anti-matter and anti-energy.

Matter consists of elementary particles, which cannot be further divided, at least with respect to current knowledge.

Anti-matter similarly consists of anti-particles.

The existence of antimatter arises from the fundamental property of symmetry of existence.

The presence of matter in the universe, implies the existence of other kind of matter, mirrored in some properties, which completes its symmetry.

In the first moments of creation of the universe, immediately after in Big Bang, Matter and Antimatter were created in equal proportions, coexisting in a sea of electromagnetic radiation.

In subsequent instances, as a result of the symmetry generated breaking down, nature favored matter, on the order of a few percentage values, such that the scales tipped toward the predominance of matter in the universe.

However, it cannot be ruled out, given the limited size of the known universe, which is about 4 percent of the existing universe, that there is abundant antimatter, but we cannot see and quantify it.

For this reason, efforts are being made to study cosmic rays, by means of space probes designed to perform detections that cannot be affected by the perturbing actions of the Earth's atmosphere, also in order to detect in the universe the natural presence, in addition to the already known positron, of other antimatter particles or better still, anti-atoms.

What is certainly known nowadays is that for every elementary particle there is the corresponding antiparticle.

For the electron, for example, there is its antielectron, called positron and $^+$, having the same mass and charge, but opposite sign.

Matter and antimatter, particles and antiparticles, have a unique feature: when they meet they vanish or rather annihilate.

The annihilation process is manifested by the formation of flashes of light (photons) following the contact of a particle with its antiparticle.

An electron in meeting a positron, they annihilate with the emission of two photons.

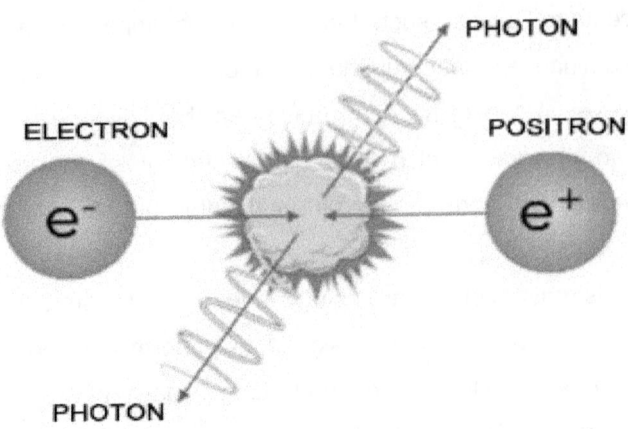

ELECTRON   PHOTON   POSITRON   PHOTON

Just think that we ourselves if we encountered our anti-I, made of antimatter, would disappear in a flash of light, actually in two.

7.2 THE ELEMENTARY PARTICLES

An elementary particle is a particle that is not further divisible, at least with reference to today's knowledge.

Elementary particles can be classified as belonging to two macro families: particles constituting matter and particles carrying forces.

The elementary particles constituting matter, are grouped into the FERMION family, a name assigned in honor of the Italian physicist Enrico Fermi.

Fermions interact with each other, not through phantom distance forces or fields of any kind, as predicted by classical mechanics, but through special force-bearing particles, devoid of mass, grouped in the family of BOSONS.

Fermions and bosons are the constituents of all the known universe.

A nonelementary particle can have fermionic behavior if it consists of an odd number of elementary fermions (quark, electron,etc.), that is, in a more general way if the overall spin results fractional anyway (1/2+1/2+1/2=3/2). The possible presence of bosons in the constitution of a compound (nonelemental) particle does not affect the fractional spin result, since the spin of bosons is of integer type (0,1,...etc.). We can say that the fermionic behavior of a compound particle is always independent of the number of bosons.

A particle composed of any number of bosons (integer spin =0,1,...etc.) still remains a boson, by the characteristic that the sum of integers still remains an integer (1+1+1+1=4)

In the following we will examine distinctly the two families of elementary particles thus identified.

With regard to particles having mass, it should be pointed out that this quantity, dimensionally, can be expressed either in kilograms or, due to the mass-energy equivalence formulated by Einstein ($E=mc^2$), in electronvolts over the speed of light in vacuum squared (eV/c^2). The electronvolt (eV) is an alternative unit of measurement for energy that is worth $1.602176565 \cdot 10^{-19}$ Joule.

Considering that $1J = 1$ Kgm $/s^{22}$ we get

$$1 \, eV = 1.602176565 \cdot 10^{-19} \frac{Kgm^2}{s^2}$$

Dividing both members by the speed of light squared c^2 =[m /s^{22}] gives

$$1 \frac{eV}{c^2} = \frac{1.602176565 \cdot 10^{-19} \frac{Kgm^2}{s^2}}{\left(299{,}792{,}458 \, \frac{m}{s}\right)^2} = 1.78 \cdot 10^{-36} Kg$$

and finally the conversion relationship

$$1 \, Kg = 5.61 \cdot 10^{35} \, \frac{eV}{c^2}$$

As an example, if a particle has a mass of 9.109×10^{-31}kg, this may be expressed as 510,977.00 eV/c^2 or rather as 511 keV/c^2 or finally as 0.511MV/c^2 , having introduced the prefixes kilo and mega.

In the following we will give the mass of the particle equivalently in kg or in eV/c^2 .

7.3 THE FERMIONS OF GENERATION I

The elementary particles of fermionic type, referring for now only to matter, that make up the universe are: the electron (e), neutrino (v), Quark UP (u) and Quark DOWN (d).

All elementary and non-elementary fermionic particles have the common peculiarity of having semi-integer spin value (1/2, 3/2, 5/2...), following the Fermi-Dirac statistic and obeying the Pauli exclusion principle.

Let us examine in detail each of the elementary particles constituting matter.

The first fermionic particle we are going to examine is precisely **"the electron (e),"** which if you remember is the dancing particle in the atomic quantum orbital.

The electron has negative charge, equal to $-1.621 \cdot 10^{-19}$ C, when measured in Colulombs. The same charge can be denoted by $Q = -1$ by considering the charge of the same electron as the reference charge or rather as the elementary charge.

The electron has spin quantum number of $s=\frac{1}{2}$, like all other elementary particles belonging to the fermion family.

Its mass is very small: its weight turns out to be 1/1836 that of the heaviest proton, and on average electrons make up only about 0.06 percent, of the weight of an atom.

Taking into account the theory of special relativity, the rest mass of an electron is about 9.109×10^{-31} kg.

The radius of the electron is about 10^{-22} meters.

Continuing, in order as reported at the beginning of the paragraph, we find the " **electron neutrino (ν)**," a particle characterized by having a very small mass, so small that it was initially thought to have none.

The name "neutrino" originated as a joking diminutive of the larger neutron.

Its mass is one hundred thousand to one million times less than that of the electron.

Being an elementary fermion, its spin quantum number is $s=\frac{1}{2}$.

Its charge is neutral, $Q = 0$, thus indifferent to electromagnetic fields, which is precisely why it is difficult to detect.

We have seen in previous chapters how it is generated in decay processes, especially as a result of nuclear fusion processes in stars, continuing undisturbed, due to the absence of charge, on its way at the speed of light.

$q=0$

Right now we are being hit by a number equal to 10 billion neutrinos per second, coming from the Sun alone in a time of 8 minutes.

Electrons and neutrinos belong to the group of Leptons, a term derived from the Greek Lepto (thin) precisely to indicate their lightness.

Continuing the description of elementary particles we find the **"up (u) quarks"** and **"down (d) quarks."**

The up and down quarks, denoted by the letters u and d, represent the components of neutrons and protons.

The mass of the up quark ranges between values of 3 to 8 times the mass of the electron. The down quark has mass with values ranging from a minimum of 8 to a maximum of 16 times the mass of the electron.

In relation to the proton, the up quark turns out to have a mass with values between 1/200 and 1/600, and the down quark between 1/100 and 1/200.

The electrical charge of the quark is fractional.

The charge of the up quark is +2/3, while the charge of the down quark is -1/3.

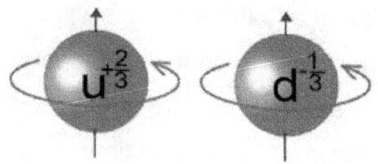

The spin quantum number is again ½.

Another quantity characterizing quarks is the "color charge," which we will better analyze in the following when we discuss gluons.

The u and d quarks, in nature, are not found in isolation, but as constituents of the neutron and proton.

A proton consists of two up quarks and one down quark, such that the electric charge of the proton formed is equal to one.

$$2\,Q_u + Q_d = Q_{proton} \Rightarrow 2\left(+\frac{2}{3}\right) - \frac{1}{3} = +1$$

A neutron, on the other hand, consists of one up quark and two down quarks, such that the electrical charge of the neutron thus constituted is zero.

$$Q_u + 2\,Q_d = Q_{proton} \Rightarrow \frac{2}{3} + 2\left(-\frac{1}{3}\right) = 0$$

7.4 SUCCESSIVE GENERATIONS OF FERMIONS

The particles examined so far belong to the first generation of fermions, which constitute the most stable part of matter and are therefore normally and easily found in nature.

In fact, the total known generations of fermions are 3.

The next two generations have higher mass and consequently higher energy; therefore, they are more unstable and more prone to rapid decay.

In fact, Generation II and III particles are produced artificially in collisions in accelerators or produced in space and detected in cosmic rays; these particles are short-lived, decaying into Generation I particles in a very short time.

With reference to the Leptons (Electron and neutrino), of Generation I, we find as Generation II: **muon neutrino (ν_μ)** and **muon (μ)**, and finally as Generation III: **tauon neutrino (ν_τ)** and **tauon (τ).**

As for Quarks (quark up and quark down), of Generation I, we find as Generation II: Quark **charm (c)** and **Quark strange (s)**, and finally as Generation III: Quark **top (t)** and **Quark bottom (b).**

The elementary particles constituting matter are thus 12, grouped into 3 generations and distinguished into quarks and leptons, all best explained by the figure below.

FERMIONS

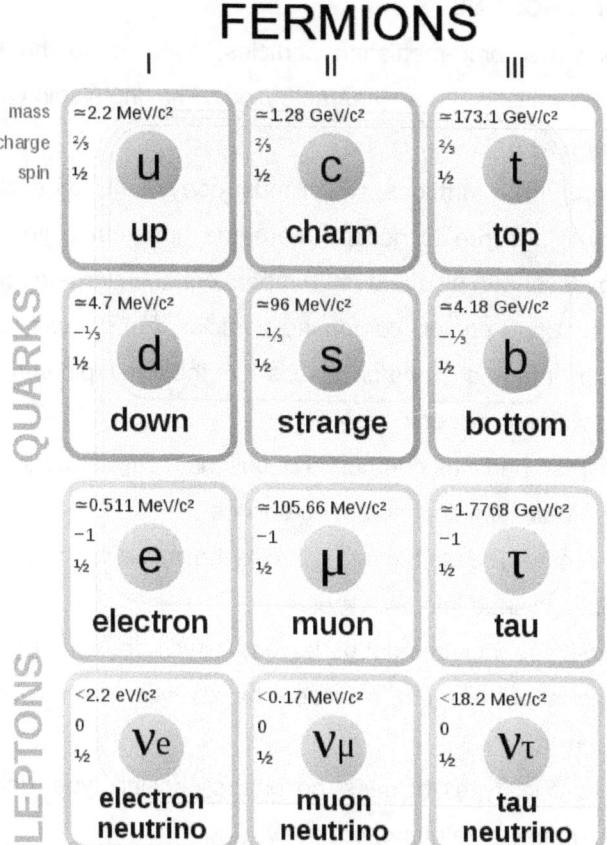

For each elementary constituent particle of matter thus identified, there is a corresponding particle of antimatter, having the same name with the prefix anti, the same mass and opposite charge. Ultimately, the elementary particles of matter and antimatter total 24.

7.5 THE BOSONS

Bosons are force-mediating particles, obedient to the Bose-Einstein statistic and are characterized by having a spin value of integer type (0, 1, 2 ...).

Bosons, unlike fermions, which must obey the Pauli exclusion principle, are free to occupy the same quantum state (same energy level with all quantum numbers equal) in large numbers. As mentioned earlier, compound particles that contain such a number of fermions and/or bosons that the total spin sum is an integer take on bosonic behavior.

In contrast, particles composed of bosons alone, always having integer total spin, continue to be bosons.

The bosonic-type elementary particles that make up the universe are distinguished into two types.

The first type includes gauge bosons, which are vector bosons and as such are characterized by direction, intensity and direction.

The second type includes bosons of scalar type, that is, representable by a numerical entity.

Gauge or vector bosons are distinguished into three types: **photon (γ), gluon (g),** and weak force **bosons (Z^0 and W).$^{\pm}$**

Instead, the scalar boson is represented by the more famous **Higgs boson**.

Let us examine each of the force-bearing elementary particles.

The first among the vector-type gauge bosons is the photon, denoted by the Greek letter γ, which represents a quantum of electromagnetic energy, and is the mediator of the electromagnetic force.

The electromagnetic force constitutes one of the four fundamental interactions known to date.

The four known fundamental forces, are distinguished as follows: electromagnetic interaction, strong nuclear interaction, weak nuclear interaction and gravitational interaction.

Electromagnetic interaction occurs through the photon and has the characteristic of having an infinite range.

The photon is massless, has zero electric charge, spin equal to 1, and is of the stable type, that is, not spontaneously decaying has an infinite average lifetime.

Since the photon has both mass and zero charge, its antiparticle is represented by the very same photon.

PHOTON

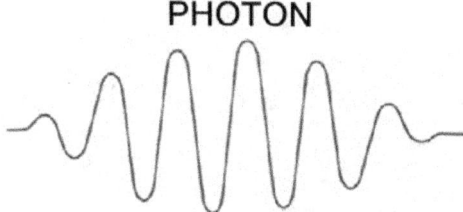

Next in the list of gauge bosons, we find the gluon, denoted by the letter g.

The gluon is the carrier of the strong nuclear interaction which, similarly to the photon, has zero mass and electric charge and as an elementary boson, spin value of 1.

The strong nuclear interaction is characterized by having a very small range, on the order of $1.4 \cdot 10^{-15}$ m, but a high intensity, hence precisely the adjective strong.

The term gluon comes from the English word "glue," given the characteristic of holding certain elementary particles together in order to form composite particles.

In particular, the gluon keeps quarks glued together, joining them into triplets to form neutrons and protons.

The proton, as seen earlier, is composed of two up quarks and one down quark, such that it has a total charge of +1, held together precisely by three gluons.

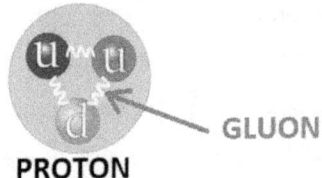

PROTON

In the neutron, however, we find one up quark and two down quarks, such that they have a total charge of zero, held together by three gluons.

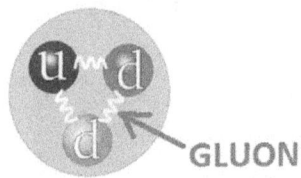

NEUTRON

In the interaction of gluons, an additional property is introduced, which is the color charge.

Color charge has nothing to do with the colors perceived by the human eye, but is a feature similar to electrical charge, best described in quantum chromodynamics (QCD).

Color charge, for example, is used to conventionally describe energy exchanges, between gluons and quarks, and is a feature of both quarks and gluons.

Quarks have only a color component, while anti-quarks, respectively, in the case of antimatter, only an anti-color component.

Gluons, on the other hand, have a mixture of two color charge components: a color and an anti-color.

Each color component is named R, G, B respectively after the name initials of the three basic colors in English: Red, Green, and Blue.

Respectively, the components of anti-color will be denoted by. $\overline{R}$ (anti-red), $\overline{G}$ (anti-green), $\overline{B}$ (anti-blue) and represented with cyan, magenta and yellow.

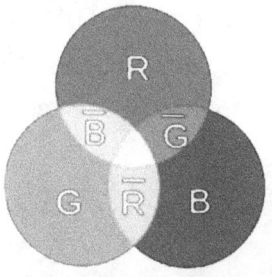

In the course of interacting with quarks, gluons, in view of the color charge they possess, in addition to keeping them bound as carriers of the strong nuclear force, also become color carriers, thus exchanging color charge with quarks.

In the proton and neutron there is a continuous exchange of color charge, by the gluons, always under the principle of conservation of total color charge, which will remain unchanged.

In analogy to what happens in classical mechanics, by rotating Newton's disk, it happens that the superposition of the three colors R, G, B will result in absence of color, that is, the color white.

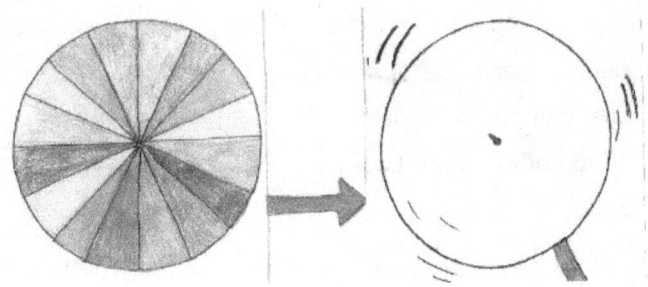

Ultimately this implies that both the proton and neutron, although made up of three quarks of color R, G, B, globally appear to have no color charge, that is, they appear white in color, which is actually the result of a continuous rainbow dance between quarks by the gluons.

Continuing our examination of additional particles belonging to the type of gauge bosons, we find the Z bosons^0 and W$^\pm$, carriers of the weak nuclear force.

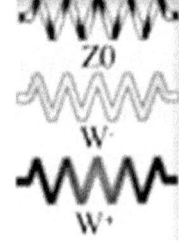

The adjective weak is a consequence of the intensity of that nuclear force, which is about 100,000 times less than the strong interaction.

The spin value, for both vector bosons, is equal to an integer and in the specific case equal to 1.

The Z bosons[0] and W±, unlike the previous bosons belonging to the same type (photon and gluon), are massive, with mass values of about 80 and 90, respectively GeV/c².

Because of their high mass, these bosons have a short average lifetime of about 3×10^{-24} seconds.

While the Z boson[0] has zero charge, the W boson± can have +1 or -1 charge, so the interaction mediated by the Z boson[0] is called "current neutral" and the interaction mediated by the W boson± is called "current charged."

In the course of the charged current interaction (mediated by the W boson±), it happens that one particle transforms (decays) into other particles with different charge.

For example, an electron, having a negative charge, can emit a W boson⁻ and become a neutrino, or it can absorb a W boson⁺ and still turn into a neutrino, as best in the following schematic.

$$e \Rightarrow W^- + v$$
$$e + W^+ \Rightarrow v$$

The weak force, through the bosons Z^0 and W± , is responsible for the phenomenon of radioactivity and in particular the beta decay of atomic nuclei associated with it.

Let us analyze, in order to better understand the presence of said bosons, the process of β⁻ decay of a neutron, previously covered with radioactivity

$$n \longrightarrow p + e^- + \overline{v_e}$$

Let us examine in detail what happens at the elementary particle level, that is, what happens inside the neutron and proton.

The neutron we have seen consists of two down quarks and one up quark, held together by gluons.

In order for a neutron to decay into a proton, it is necessary for a down quark to transform into an up quark,

PROTON   NEUTRON

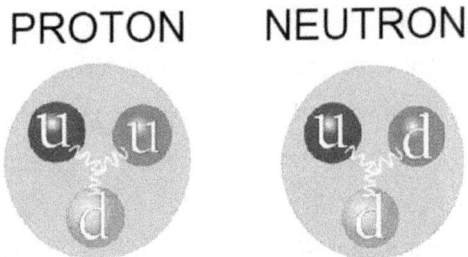

Given that the d quark has charge has charge *-1/3* and the u quark has charge *+2/3*, balancing the equation will require the emission of a *W* boson⁻ , which has negative charge equal to *-1*, such that -1/3 - (-1) = 2/3.

In terms of elementary particles

$$d \longrightarrow u + W^-$$

The neutron decay process is best illustrated graphically as follows.

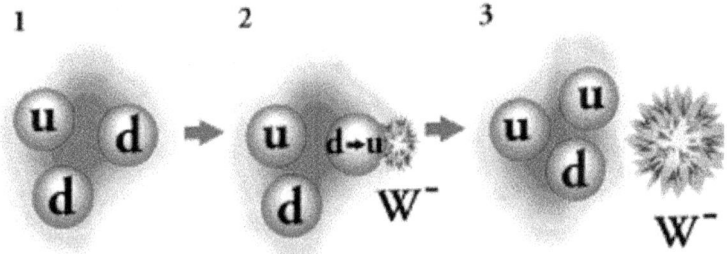

The W- boson, as mentioned earlier, however, is short-lived and decays into an electron and an anti-neutrino

$$W^- \longrightarrow e^- + \overline{\nu}_e$$

protone

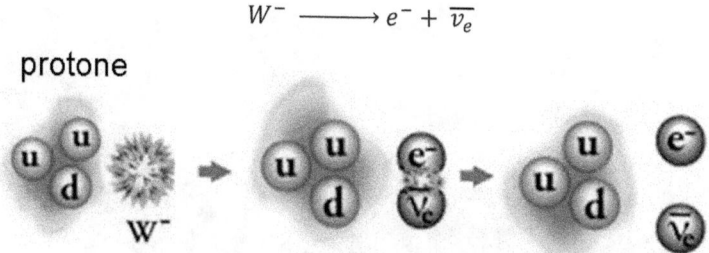

Ultimately, the overall decay reaction remains unchanged from what was seen in the previous paragraphs, remembering, however, that the same reaction hides an intermediate decay with the presence of the W boson$^-$.

$$n \longrightarrow p + e^- + \overline{\nu}_e$$

By way of summary, the following table summarizes the type of known bosons, distinguished into vector or gauge bosons and scalar bosons.

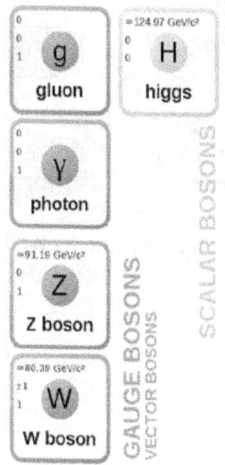

The Higgs Boson, qualde boson of scalar type, will be appropriately treated in the following section.

COMMONS.WIKIMEDIA.ORG

"It shouldn't be a Higgs field. If it's anybody's, it should be Goldstone field, I think. When Nambu wrote his short paper in 1960, Jeffrey Goldstone of Cambridge University, who was visiting Cern, heard about it. He then wrote a paper which was conceptually similar to what Nambu had done, but a simpler model."

PETER HIGGS
https://www.brainyquote.com/quotes/

7.6 THE HIGGS BOSON

The Higgs boson is a scalar-type boson, very massive, with zero charge, zero integer spin; like the case of the photon, its antiparticle is equal to its own particle.

This boson being of scalar type, unlike other vector bosons, is not force mediator but is mass mediator.

For this reason, the Higgs boson is responsible for the mass of all elementary particles.

It was named in honor of British physicist **Peter Ware Higgs,** who solved theoretically in 1964 the problem concerning the origin of the constitution of mass in elementary particles by introducing theoretically, a complex  scalar field and a new particle: the Higgs field and boson.

The Higgs field is a complex scalar field that in the instants after  the Big Bang, in terms of billionths of a second, instantaneously permeated space.

At such instants, the existing particles, originally massless, interacted with this scalar-like field through the mediation of the associated "quantum" that is precisely the Higgs boson. From such interactions, however, no forces of any kind arose, but a transfer of energy occurred.

Due to mass-energy equivalence, energy transfer initially conferred mass on W-type gauge bosons$^{\pm}$ and Z-type bosons^{0}, while the photon and gluon remained massless.

Later mass was also conferred on fermions (quarks and leptons). The conferring of mass on said elementary particles, caused them to slow down, as, by the theory of special relativity, they were inhibited from being able to continue traveling at the speed of light.

The boson as predicted theoretically found experimental confirmation through its observation in CERN's LHC particle accelerator by the ATLAS and CMS Experiments.

In an announcement made on July 4, 2012, at a conference held in the CERN auditorium, in the presence of Peter Higgs, the discovery of a particle compatible with the Higgs boson was announced, whose mass was experimentally found to be about 126.5 GeV/c^2 - 125.3 GeV/c^2.

That discovery led the international scientific community to award Peter Higgs the Nobel Prize in Physics in 2013.

The Higgs boson is also known as the "God particle," the name of which comes from the publisher's change of the original nickname "Goddamn particle" from the title of a popular physics book by Leon Lederman.

Regarding this appellation, Higgs said he disagreed with the expression, finding it potentially offensive to people of religious faith.

COMMONS.WIKIMEDIA.ORG

I was an embarrassment to the department when they did research assessment exercises. A message would go round the department: 'Please give a list of your recent publications.' And I would send back a statement: 'None.'

PETER HIGGS
https://www.brainyquote.com/quotes/

7.7 THE GRAVITON BOSON

The elementary particles known to date and according to the standard model classification are summarized in the table below:

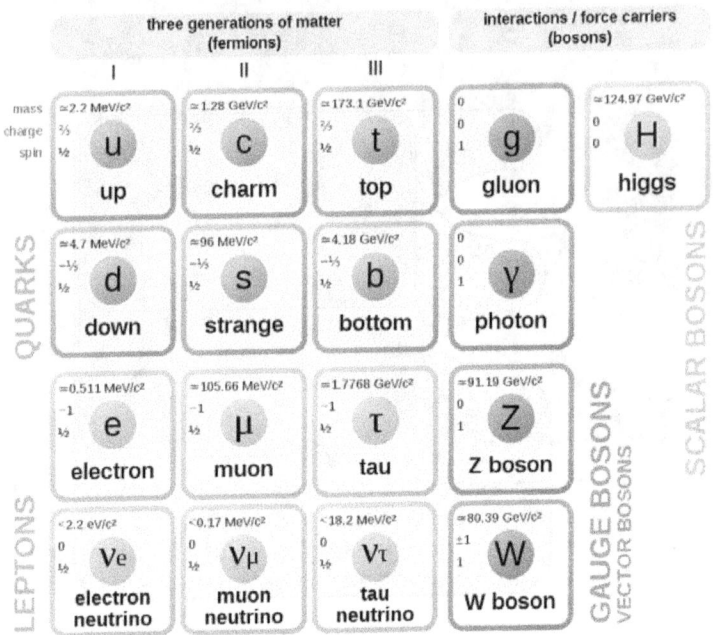

Left out of the previous classification is an additional boson called the "graviton," having zero mass, zero charge, spin equal to 2 and infinite range.

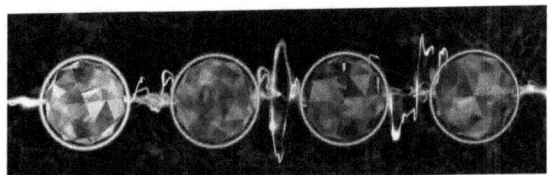

Actually said graviton boson to date is only a hypothesis having not yet had any experimental evidence.

His research is based on an attempt to unite gravitational theory with the theory of quantum mechanics.

In fact, the graviton is supposed to be responsible for transmitting the force of gravity.

Thus the graviton would go to mediate the force of gravity of the attractive type between two bodies placed at any distance, through continuous exchanges in the limit from the speed of light as required by the theory of special relativity, unlike what happens in classical physics with the formation of the gravitational field.

The main issue is inherent in the detection of the graviton, since such a particle, if it exists, would have a very weak level of interaction.

8 PARTICLE ACCELERATORS

Elementary particles and more can be observed through cosmic rays (improperly called "rays") coming from the universe and directed at the Earth's surface.

The universe in this sense is a hotbed of elementary particles.

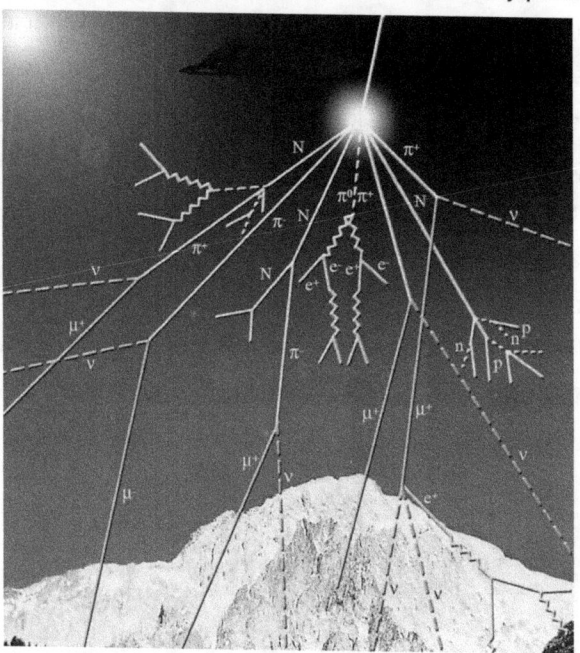

The limitation of observing cosmic rays lies in the interaction of the constituent particles with the layers of Earth's atmosphere, which is why efforts are made to study said cosmic rays beyond Earth's atmosphere by sending appropriate space probes.

There is another way of artificially searching for elementary particles.

However, for this purpose it is necessary to create high energy values in order to simulate what happens in the universe and cosmic rays.

It is possible to create high values of kinetic energy by acting on the speed of particles, accelerating them in special "accelerators" of both circular and linear types.

In a linear accelerator (LINAC), particles are accelerated along a straight path against a fixed target. Linear accelerators are very common, for example, a cathode ray tube is a linear electron accelerator. These accelerators are also used to provide the initial energy to particles that will later be fed into more powerful circular accelerators. The longest linear accelerator in the world is the Stanford Linear Accelerator, which is 3 kilometers long.

Circular accelerators have a toroidal shape.

In such accelerators, by appropriately confining the original particles input with electromagnetic fields, due to the possibility of periodic motion, high velocities can be achieved by proceeding to continuous acceleration.

After the particles gain velocity and thus appropriate energy, we proceed to cause them to collide.

From said collision of highly energetic particles, something strange happens: by mass-energy equivalence, the particles turn into other kinds of particles.

It is as if by colliding 2 pears at high speeds, these result in a banana, an apple and an orange.

Only, it is not so easy to read the results inside a particle accelerator; in fact as a result of the collisions between such highly energetic particles, deriving useful information from the results is like putting together the pieces of an object thrown from a skyscraper.

The largest existing accelerator in the world is the LHC (large

hadron collider) built inside a 27-kilometer-long, circular underground tunnel located at an average depth of 100 meters (330 feet) on the border between France and Switzerland at CERN in Geneva.

This accelerator can accelerate hadrons, which are nonelementary subatomic particles consisting of quarks also associated with antiquarks, such as protons and heavy ions.

It succeeds in making said particles reach a speed of up to 99.9999991% of the speed of light and subsequently make them collide, with an energy that in May 2015 reached 13 teraelectronvolt (TeV), very close to the machine's theoretical limit of 14 TeV.

The machine operates under vacuum conditions, accelerating through more than 1,600 superconducting magnets that realize a magnetic field of about 8 Tesla, which is necessary to keep two beams of particles circulating in opposite directions in orbit at the expected energy.

The collision is allowed to occur in appropriate detectors, called detectors, where post-collision observation takes place.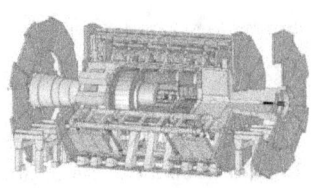

The detectors consist of several concentric cylindrical layers suitable for observing both charged and neutral particles, and both massive and massless particles, through charge detectors, calorimeters for measuring particle energy, spectrometers and magnet systems.

The only particles that cannot be detected are neutrinos, due to their characteristic of having a very small mass associated with the absence of electric charge.

At one time, the traces of particles generated as a result of the collision were observed in special bubble chambers, first devised and built by U.S. physicist and neurobiologist Donald Arthur Glaser in 1952, the discovery of which earned him the Nobel Prize in Physics in 1960.

The bubble chamber represented an evolution of the older fog chamber as an instrument for detecting elementary particles devised by British physicist Charles Thomson Rees Wilson in 1899 and later perfected in 1912.

The fog chamber consists of an airtight box that contains air supersaturated with water vapor, which upon the passage of any electrically charged particle causes ionization of the atoms with which it collides, consequently creating along its path a trail of ionized atoms around which the supersaturated vapor collects to form tiny droplets.

The trace left by the trajectory traversed by the particle can be photographed through a transparent wall of the box, and from this it can be traced, with special precautions, to the determination of the characteristics and nature of the particle.

The bubble chamber, on the other hand, consists of a cylindrical metal vessel containing a superheated and compressed liquid, thus in a metastable condition.

In such a case, a fast, charged particle passing through the vessel ionizes the atoms of the liquid and at the same time slows down its travel, losing energy as a result of the collisions.

Along the path of the particle, positive and negative ions are created around which the liquid begins to boil, thus leaving a trace of the passage.

By taking several photos from different angles, a stereoscopic spatial reconstruction of the tracks is obtained.

Since the bubble chamber consists of liquid, thus at a higher density than the fog chamber, greater ionization is achieved resulting in better trace definition and at the same time better braking action useful for observing light or low-energy particles.

Various types of fog or bubble chambers are still being made to this day for educational use because of the striking images that can be obtained.

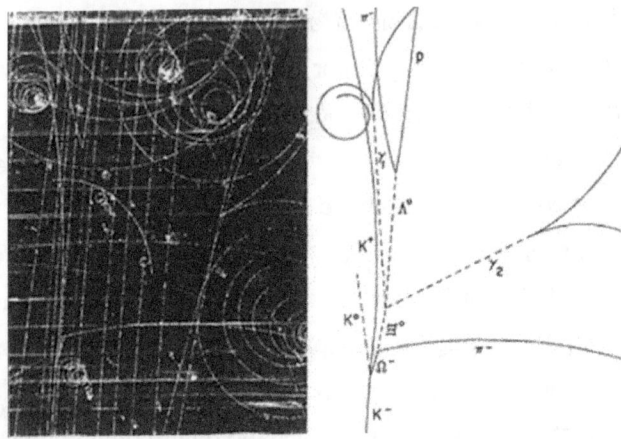

Otherwise for professional purposes today particles are detected with special superconductors and the traces reconstructed digitally.

BIBLIOGRAPHY

Fenomeni radioattivi, dai nuclei alle stelle - Giorgio Bendiscioli - Springer Vergal Italia 2013

I quanti e la vita – Niels Bohr - Universale scientifica Boringhieri – Prima edizione 1965 – Ristampe 1969, 1974

Teoria dei quanti – John Polkinghorne – Codice edizioni Torino - 2007

Meccanica quantistica, il minimo indispensabile per fare della (buona) fisica – Leonard Susskind Art Friedman – Raffaello Cortina editore - 2015

Dalla fisica classica alla fisica quantistica – Carlo Tarsitani – Editori riuniti university press – 2009

L'esperimento più bello – Giorgio Lulli – Apogeo – 2013

I principi della meccanica quantistica – Paul Adrien M. Dirac – Bollati Boringhieri editore Torino – prima edizione 1959, ristampa 2014

Il bizzarro mondo dei quanti – Silvia Arrayo Camejo – Springer - 2012

L'atomo e le particelle elementari – Massimo Teodorani – Macro Edizioni – prima edizione 2007, ristampa 2012

Il mondo secondo la fisica quantistica – Fabio Fracas – Sperling & Kupfer - 2017

Bibliografia e immagini da Web:

Immagini
commons.wikimedia.org

Joseph John Thomson
https://it.wikipedia.org/wiki/Joseph_John_Thomson
https://it.wikipedia.org/wiki/Modello_atomico_di_Thomson

Ernest Rutherford
https://it.wikipedia.org/wiki/Ernest_Rutherford
https://it.wikipedia.org/wiki/Esperimento_di_Rutherford

Max Planck
https://it.wikipedia.org/wiki/Max_Planck
https://it.wikipedia.org/wiki/Catastrofe_ultravioletta
https://it.wikipedia.org/wiki/Corpo_nero
https://it.wikipedia.org/wiki/Costante_di_Planck
https://it.wikipedia.org/wiki/Spettro_elettromagnetico

Niels Bohr
https://it.wikipedia.org/wiki/Niels_Bohr
https://it.wikipedia.org/wiki/Modello_atomico_di_Bohr

Arnold Sommerfeld
https://it.wikipedia.org/wiki/Arnold_Sommerfeld
https://it.wikipedia.org/wiki/Formula_di_Wilson-Sommerfeld

Orbitale atomico
https://it.wikipedia.org/wiki/Orbitale_atomico

Stato quantistico di Spin
https://it.wikipedia.org/wiki/Spin

Esperimento di Stern-Gerlach
https:// it.wikipedia.org/wiki/Esperimento_di_Stern-Gerlach

Wolfgang Pauli
https://it.wikipedia.org/wiki/Wolfgang_Pauli
https://it.wikipedia.org/wiki/Principio_di_esclusione_di_Pauli

Werner Karl Heisenberg
https://it.wikipedia.org/wiki/Werner_Karl_Heisenberg
https://it.wikipedia.org/wiki/Principio_di_indeterminazione_di_Heisenberg

Erwin Schrödinger
https://it.wikipedia.org/wiki/Erwin_Schr%C3%B6dinger
https://it.wikipedia.org/wiki/Equazione_di_Schr%C3%B6dinger
https://it.wikipedia.org/wiki/Funzione_d%27onda
https://it.wikipedia.org/wiki/Paradosso_del_gatto_di_Schr%C3%B6dinger

Louis-Victor Pierre Raymond de Broglie
https://it.wikipedia.org/wiki/Louis-Victor_Pierre_Raymond_de_Broglie
https://it.wikipedia.org/wiki/Ipotesi_di_de_Broglie

Paul Dirac
https://it.wikipedia.org/wiki/Paul_Dirac
https://it.wikipedia.org/wiki/Notazione_bra-ket

Thomas Young
https://it.wikipedia.org/wiki/Thomas_Young
https://it.wikipedia.org/wiki/Esperimento_di_Young

Alain Aspect
https://it.wikipedia.org/wiki/Alain_Aspect

Joh n Stewart Bell
https://it.wikipedia.org/wiki/Teorema_di_Bell

James Chadwick
https://it.wikipedia.org/wiki/James_Chadwick
https://it.wikipedia.org/wiki/Neutrone

Isotopi

https://it.wikipedia.org/wiki/Isotopi_dell%27idrogeno
Radioattività
https://it.wikipedia.org/wiki/Radioattivit%C3%A0
https://it.wikipedia.org/wiki/Decadimento_alfa
https://it.wikipedia.org/wiki/Decadimento_beta
https://it.wikipedia.org/wiki/Raggi_gamma
https://it.wikipedia.org/wiki/Radiazioni_ionizzanti
https://it.wikipedia.org/wiki/Metodo_del_carbonio-14

Fissione e fusione nucleare
https://it.wikipedia.org/wiki/Fissione_nucleare
https://it.wikipedia.org/wiki/Fusione_nucleare
https://it.wikipedia.org/wiki/Bomba_all%27idrogeno
https://it.wikipedia.org/wiki/Reattore_nucleare_a_fusione
https://it.wikipedia.org/wiki/Reattore_nucleare_a_fissione
https://it.wikipedia.org/wiki/Bomba_atomica
https://it.wikipedia.org/wiki/Nucleosintesi_stellare
https://it.wikipedia.org/wiki/Pulsar
https://it.wikipedia.org/wiki/Quasar
https://it. wikipedia.org/wiki/Buco_nero

Antimateria
https://it.wikipedia.org/wiki/Antimateria

Le particelle elementari
https://it.wikipedia.org/wiki/Particella_elementare
https://it.wikipedia.org/wiki/Fermione
https://it.wikipedia.org/wiki/Bosone_(fisica)
https://it.wikipedia.org/wiki/Quark_(particella)

Peter Higgs
https://it.wikipedia.org/wiki/Peter_Higgs
https://it.wikipedia.org/wiki/Bosone_di_Higgs

Gravitone
https://it.wikipedia.org/wiki/Gravitone

Raggi cosmici
https://it.wikipedia.org/wiki/Raggi_cosmici

Acceleratore di particelle
https://it.wikipedia.org/wiki/Acceleratore_di_particelle
https://it.wikipedia.org/wiki/CERN
https://it.wikipedia.org/wiki/Large_Hadron_Collider
https://it.wikipedia.org/wiki/Camera_a_nebbia
https://it.wikipedia.org/wiki/Camera_a_bolle

Il teletrasporto quantistico compie vent'anni
http://www.lescienze.it/news/2017/12/16/news/vent_anni_di_esperimenti_sul_t
eletrasporto_quantistico-3793007/